Lecture Notes in Computer Science 9611

Commenced Publication in 1973
Founding and Former Series Editors:
Gerhard Goos, Juris Hartmanis, and Jan van Leeuwen

More information about this series at http://www.springer.com/series/7407

Tomáš Kozubek · Radim Blaheta
Jakub Šístek · Miroslav Rozložník
Martin Čermák (Eds.)

High Performance Computing in Science and Engineering

Second International Conference, HPCSE 2015
Soláň, Czech Republic, May 25–28, 2015
Revised Selected Papers

 Springer

Editors
Tomáš Kozubek
VSB - Technical University of Ostrava
Ostrava
Czech Republic

Radim Blaheta
Institute of Geonics of the CAS
Ostrava
Czech Republic

Jakub Šístek
Institute of Mathematics of the CAS
Prague
Czech Republic

Miroslav Rozložník
Institute of Computer Science of the CAS
Prague
Czech Republic

Martin Čermák
VSB - Technical University of Ostrava
Ostrava
Czech Republic

ISSN 0302-9743   ISSN 1611-3349 (electronic)
Lecture Notes in Computer Science
ISBN 978-3-319-40360-1   ISBN 978-3-319-40361-8   (eBook)
DOI 10.1007/978-3-319-40361-8

Library of Congress Control Number: 2016940813

LNCS Sublibrary: SL1 – Theoretical Computer Science and General Issues

This Springer imprint is published by Springer Nature
The registered company is Springer International Publishing AG Switzerland

Preface

This volume comprises the proceedings of the Second International Conference on High Performance Computing in Science and Engineering, HPCSE 2015, which was held at the Hotel Soláň in the heart of the Beskydy Mountains, Czech Republic, from 25th – 28th May, 2015. The biennial conference was organized by IT4Innovations National Supercomputing Center at the VSB-Technical University of Ostrava and its goal was to bring together specialists in applied mathematics, numerical methods, and parallel computing to share experience and initiate new research collaborations. We are pleased that our invitation was accepted by distinguished experts from leading international research institutions.

The conference has become an international forum for exchanging ideas among researchers involved in scientific and parallel computing, including theory and applications as well as applied and computational mathematics. The focus of HPCSE 2015 was on models, algorithms, and software tools that facilitate efficient and convenient utilization of modern parallel and distributed computing architectures, as well as on large-scale applications.

The Scientific Committee of HPCSE 2015 comprised Radim Blaheta, Zdeněk Dostál, Tomáš Kozubek, Miroslav Tůma, Jakub Šístek, Zdeněk Strakoš, and Vít Vondrák.

The plenary talks were presented by:

- Owe Axelsson from the Institute of Geonics of the CAS (Czech Republic)
- Cevdet Aykanat from Bilkent University, Ankara (Turkey)
- Marc Baboulin from Inria, Paris (France)
- Santiago Badia from CIMNE, Barcelona (Spain)
- Jed Brown from the Argonne National Laboratory, Illinois (USA)
- Fehmi Cirak from the University of Cambridge (UK)
- Jacek Gondzio from the University of Edinburgh (UK)
- Akhtar A. Khan from Rochester Institute of Technology, Rochester (USA)
- Johannes Kraus from the University of Essen (Germany)
- Jaroslav Kruis from the Czech Technical University in Prague (Czech Republic)
- Julien Langou from the University of Colorado, Denver (USA)
- Jan Mandel from the University of Colorado, Denver (USA)
- Dan Negrut from the University of Wisconsin-Madison, Madison (USA)
- Ulrich Rüde from the University of Erlangen (Germany)
- Valeria Simoncini from the University of Bologna (Italy)
- Wim Vanroose from the University of Antwerp (Belgium)
- Barbara Wohlmuth from the Technical University of Munich (Germany)
- Roman Wyrzykowski from Czestochowa University of Technology (Poland)
- Walter Zulehner from Johannes Kepler University, Linz (Austria)

The conference was supported by IT4Innovations National Supercomputing Center and the project OP EC "New Creative Teams in Priorities of Scientific Research" no. CZ.1.07/2.3.00/30.0055. This project was funded by structural funds of the European Union and the State Budget of the Czech Republic. It is our pleasure to acknowledge this support.

HPCSE 2015 was a fruitful event that presented interesting lectures, featuring new ideas as well as the beauty of applied mathematics, numerical linear algebra, optimization methods, and high performance computing; the conference facilitated the formation of new or the strengthening of the existing collaborations and friendships.

This meeting attracted more than 100 participants from 11 countries. All participants were invited to submit an original paper to this book of proceedings. We thank for all the contributors and the reviewers of the papers for their work. We hope that readers will find this volume useful, and we would like to cordially invite them to participate at the next conference, HPCSE 2017, which is planned to be held in the same place from 22nd – 25th May, 2017.

The proceedings have been edited by Tomáš Kozubek, Radim Blaheta, Jakub Šístek, Miroslav Rozložník, and Martin Čermák.

May 2016   Tomáš Kozubek

Organization

Program Committee

Radim Blaheta	Institute of Geonics of the CAS, Czech Republic
Zdeněk Dostál	IT4Innovations, VŠB - Technical University of Ostrava, Czech Republic
Tomáš Kozubek	IT4Innovations, VŠB - Technical University of Ostrava, Czech Republic
Zdeněk Strakoš	Charles University in Prague, Czech Republic
Jakub Šístek	Institute of Mathematics of the CAS, Czech Republic
Miroslav Tůma	Charles University in Prague, Czech Republic
Vít Vondrák	IT4Innovations, VŠB - Technical University of Ostrava, Czech Republic

Contents

Large Scale Lattice Boltzmann Simulation for the Coupling of Free and Porous Media Flow

Ehsan Fattahi, Christian Waluga, Barbara Wohlmuth, and Ulrich Rüde$^{(\boxtimes)}$

Fakultät für Mathematik, Technische Universität München, Munich, Germany
{fattahi,waluga,wohlmuth}@ma.tum.de, ulrich.ruede@fau.de
http://www-m2.ma.tum.de

Abstract. In this work, we investigate the interaction of free and porous media flow by large scale lattice Boltzmann simulations. We study the transport phenomena at the porous interface on multiple scales, i.e., we consider both, computationally generated pore-scale geometries and homogenized models at a macroscopic scale. The pore-scale results are compared to those obtained by using different transmission models. Two-domain approaches with sharp interface conditions, e.g., of Beavers–Joseph–Saffman type, as well as a single-domain approach with a porosity depending viscosity are taken into account. For the pore-scale simulations, we use a highly scalable scheme with a robust second order boundary handling. We comment on computational aspects of the pore-scale simulation and on how to generate pore-scale geometries. The two-domain approaches depend sensitively on the choice of the exact position of the interface, whereas a well-designed single-domain approach can lead to a significantly better recovery of the averaged pore-scale results.

Keywords: Lattice Boltzmann method · Pore-scale simulation · Two domain approach · Darcy Navier-Stokes coupling · Interface conditions

1 Introduction

Transport phenomena in porous materials are important in many scientific and engineering applications such as catalysis, hydrology, tissue engineering and enhanced oil recovery. In the past several decades, flow in porous media has been studied extensively both experimentally and theoretically. We refer the interested reader to the textbook [1] and the references therein. In porous media flow, we usually distinguish between three scales: the pore-scale, the representative elementary volume (REV) scale and the domain scale. The REV is defined as the minimal element for which macroscopic characteristics of a porous flow can be observed. Because experimental setups for many practical questions may be too expensive or even impossible to realize, numerical simulation of porous media flow can be a useful complementary method to conventional experiments.

To describe the flow in the bulk of the porous medium, Darcy's law is commonly used in the form

$$\mu \mathbf{K}^{-1} \mathbf{u} = \mathbf{F} - \nabla p, \tag{1}$$

© Springer International Publishing Switzerland 2016
T. Kozubek et al. (Eds.): HPCSE 2015, LNCS 9611, pp. 1–18, 2016.
DOI: 10.1007/978-3-319-40361-8_1

where μ is the dynamic viscosity of the fluid, $\mathbf{K}$ is the permeability tensor of the porous medium, $\mathbf{F}$ is the body force, and $\mathbf{u}$ and p are averaged velocity and pressure quantities, respectively. However, when a porous medium and a free flow domain co-exist, e.g., in a river bed, there is no uniquely accepted model for the transition between the Darcy model and the free flow. Different approaches based on two-domain or on single-domain models are available. Using a single-domain in combination with the Brinkman equation that modifies Darcy's law by a viscous term

$$- \mu_{\text{eff}} \nabla^2 \mathbf{u} + \mu \mathbf{K}^{-1} \mathbf{u} = \mathbf{F} - \nabla p, \tag{Br}$$

allows to model a smooth transition (see e.g. [2–4]). Here μ_{eff} is an effective dynamic viscosity in the porous region. However, determining appropriate viscosity parameters for the Brinkman model in the transient region is challenging [4–6]. Furthermore, the penetration of the flow into the porous medium is found to depend on the roughness coefficient of the surface; see e.g. [7–10].

Alternatively, one can use a two-domain approach in combination with a sharp interface transmission condition. Considering the (Navier-)Stokes equation in the free flow region and the Brinkman (or Darcy) equation in the porous region, the interface plays an important role. Proceeding from the experimental investigation of Poiseuille flow over a porous medium, Beavers and Joseph [11] introduced an empirical approach that agreed well with their experiment; see also [3]. They suggested to use a slip-flow condition at the interface, i.e., the velocity gradient on the fluid side of the interface is proportional to the slip velocity. For simplicity, we consider a domain for which the interface is aligned with the flow direction. The Beavers–Joseph relation is formulated as

$$\left. \frac{dU}{dz} \right|_{z=0^+} = \frac{\alpha}{\sqrt{k}} \left(U_s - U_m \right), \tag{BJ}$$

where z denotes the coordinate perpendicular to the interface, $U = U(z)$ is the mean velocity in flow direction, U_s is the slip velocity at the interface $z = 0^+$, U_m is the seepage velocity that is evaluated far from the plane $z = 0$ in the porous region, k is the permeability, and α is a phenomenological dimensionless parameter, only depending on the porous media properties that characterize the structure of the permeable material within the boundary region which typically varies between 0.01 and 5 [12,13]. We refer to [14,15] and the references therein for the interface coupling of two-phase compositional porous-media flow and one-phase compositional free flow.

In 1971, Saffman [16] found that the tangential interface velocity is proportional to the shear stress. He proposed a modification of the BJ condition as

$$\left. \frac{\sqrt{k}}{\alpha} \frac{dU}{dz} \right|_{z=0^+} = U_s + O(k). \tag{BJS}$$

More than two decades later, Ochoa-Tapia and Whitaker [17] proposed an alternative modification of the BJ condition which includes the velocity gradient on both sides of the interface as

$$\mu_{\text{eff}} \frac{dU}{dz}\bigg|_{z=0^-} - \mu \frac{dU}{dz}\bigg|_{z=0^+} = \frac{\mu}{\sqrt{k}} \beta U_s. \qquad \text{(OTW)}$$

Here the jump-coefficient β is a free fitting parameter that needs to be determined experimentally [18]. Different expressions for the effective viscosity μ_{eff} can be found in the literature. For instance, Lundgren [19] suggested a relation of the form $\mu_{\text{eff}} = \mu/\epsilon$, where ϵ is the porosity.

All of the interface conditions mentioned above require the a priori knowledge of the exact position of the interface [20–22], which is for realistic porous geometries often not the case. Additionally both, single-domain and two-domain, homogenized models rely on assumptions whose validity is not automatically guaranteed and depend on additional parameters. Traditional experiments to validate and calibrate such models are often costly, time consuming and difficult to set up. On the other hand, modern high performance computers enable the development of increasingly complex and accurate computational models resolving pore-scale features. Designing highly efficient solvers for partial differential equations is one of the challenges of extreme scale computing. While finite volume/clement/difference schemes give rise to huge algebraic systems, lattice Boltzmann methods are intrinsically parallel and attractive from the computational complexity point of view. Thus fully resolved direct numerical simulation based on first principles modeling is not only feasible nowadays but also provides an attractive possibility for validation and calibration. The LBM is commonly used as a tool to investigate the small scale phenomena in porous media. Prior work studied dense packing of spherical [25] and non-spherical particles [26,30] and showed very good agreement between the numerical and experimental results.

As a next step in this direction, we here carry out a direct numerical simulation of free flow over a porous medium. The model porous media geometry is constructed by generating a random sphere-packing using a parallel in-house multi-body simulation framework called PE [23]. In the pore geometries constructed such, the flow equations are solved with full geometrical resolution. This naturally leads to high computational cost requiring the use of high end parallel computing. As we will show by performance analysis, the in-house lattice Boltzmann solver WALBERLA [24] exhibits excellent performance and parallel scalability for these pore-scale simulations.

We use the results of the direct numerical simulation of flow over and through the porous media as reference solution and evaluate several sharp-interface conditions. As a further example, we also use a homogenized lattice Boltzmann model as a REV scale simulation and show the capability of this model to reproduce the pore-scale results with high accuracy.

2 Numerical Method

The lattice Boltzmann method (LBM) has been successfully applied to simulate porous media flow [27–30]. The kinetic nature of the LBM enables it for fluid systems involving microscopic interactions, e.g., flow through porous media. Furthermore, its computational simplicity, its amenability to a simple and efficient implementation and parallelization, and its ability to handle geometrically complex domains makes it an applicable tool to simulate porous media flow on the pore-scale.

The LBM can also be applied to model the fluid flow in porous media at the REV scale. The most commonly used models are the Darcy, the Brinkman-extended Darcy and the Forchheimer-extended Darcy models. This last approach accounts for the flow resistance in the standard LBM by modifying the body-force or equilibrium terms, leading to the recovery of either Darcy-Brinkmans equations or generalized Navier-Stokes equations [31–33]. The general model of porous media flow should consider the fluid forces and the solid drag force in the momentum equation [34]. Guo and Zhao [35] proposed a model to include the porosity into the equilibrium distribution and added a force term to the evolution equation to account for drag forces of the medium. The non-linear inertial term is not included in the Brinkman model either, and thus, this model is only suitable for low-speed flow. In this approach, the detailed structure of the medium is ignored, and the statistical properties of the medium are included to represent the porous effects.

2.1   The Lattice Boltzmann Equation

The LBM originates historically from the lattice-gas automata method and can also be viewed as a special discrete scheme for the Boltzmann equation with discrete velocities

$$\mathbf{f}(\mathbf{x} + \mathbf{e}_k \Delta t, t + \Delta t) - \mathbf{f}(\mathbf{x}, t) = \mathbf{\Omega}(\mathbf{x}, t) + F_k \Delta t \tag{2}$$

where $\mathbf{e}_k$ is the particle velocity, and $\mathbf{\Omega}(\mathbf{x}, t)$ is the collision operator. For the three dimensional lattice model D_3Q_{19}, $\mathbf{f}(\mathbf{x}, t) = (f_0(\mathbf{x}, t), f_1(\mathbf{x}, t), ..., f_{18}(\mathbf{x}, t))^T$ is a 19-dimensional vector of distribution functions. F_k is the force that acts as a source term to drive the flow.

A common approach is to use the Bhatnagar-Gross-Krook (BGK) [36] model that features a single-relaxation-time (SRT) approximation for the collision operator. However, it has been shown that using the SRT leads to a nonphysical viscosity dependence of boundary locations and also suffers from poor stability properties [37,38]. Here, we use the TRT collision operator in which the relaxation time of the symmetric and anti-symmetric components of the distribution function are separated. For an in-depth discussion of the TRT model, we refer to [39–41]. As proposed by Ginzburg [39], the TRT model uses two relaxation rates ω^+ and ω^- where ω^+ is used for even order moments, and ω^- is used for odd order moments

$$\mathbf{\Omega}(\mathbf{x}, t) = -\omega^+ \left(\mathbf{f}^+(\mathbf{x}, t) - \mathbf{f}^{eq,+}(\mathbf{x}, t) \right) - \omega^- \left(\mathbf{f}^-(\mathbf{x}, t) - \mathbf{f}^{eq,-}(\mathbf{x}, t) \right), \tag{3}$$

and

$$f_k^+ = \frac{f_k + f_{\bar{k}}}{2} \quad , \quad f_k^- = \frac{f_k - f_{\bar{k}}}{2}. \tag{4}$$

Here $\bar{k}$ denotes the opposite direction of the index k in the velocity set. The first eigenvalue is related to the kinematic viscosity as $1/\omega^+ = 3\nu + 0.5$, and the second eigenvalue ω^- controls the anti-symmetric modes which do not enter in the second order mass and momentum conservation equations, hence, it can be assumed as a free parameter. Due to stability reasons, ω^- has to be selected in $(0, 2)$ [39]. The equilibrium distribution function $\mathbf{f}^{\text{eq}}(\mathbf{x}, t_n)$ for incompressible flow is given by [42]

$$f_k^{\text{eq}}(\mathbf{x}, t_n) = w_k \left\{ \delta\rho + \rho_0 \left[c_s^{-2}\mathbf{e}_k \cdot \mathbf{u} + \tfrac{1}{2}c_s^{-4}(\mathbf{e}_k \cdot \mathbf{u})^2 - \tfrac{1}{2}c_s^{-2}\mathbf{u} \cdot \mathbf{u} \right] \right\}, \quad (5)$$

where w_k is a set of weights normalized to unity, $\rho = \rho_0 + \delta\rho$. Here $\delta\rho$ is the density fluctuation, and ρ_0 is the mean density which we set to $\rho_0 = 1$. $c_s = \Delta x/(\sqrt{3}\Delta t)$ is the lattice speed of sound, while Δx denotes the lattice cell width. The macroscopic values of density ρ and velocity $\mathbf{u}$ can be calculated from $\mathbf{f}$ as zeroth and first order moments with respect to the particle velocity, i.e.,

$$\rho = \sum_{k=0}^{18} f_k, \qquad \mathbf{u} = \rho_0^{-1} \sum_{k=0}^{18} \mathbf{e}_k f_k. \quad (6)$$

In a lattice Boltzmann scheme, we typically split the computation into a collision and a streaming step that are given as

$$\tilde{f}_k(\mathbf{x}, t_n) - f_k(\mathbf{x}, t_n) = \Omega(\mathbf{x}, t) + F_k \Delta t, \qquad \text{(collision)}$$

$$f_k(\mathbf{x} + \mathbf{e}_k \Delta t, t_{n+1}) = \tilde{f}_k(\mathbf{x}, t_n), \qquad \text{(streaming)}$$

respectively, for $k = 0, \ldots, 18$. The execution order of these two steps is arbitrary and may vary from code to code for implementation reasons.

In addition, for linear steady flow, it has been demonstrated [40] that most of the macroscopic errors/quantities of the TRT depend on $\Lambda = \left(\frac{1}{\omega^+} - \frac{1}{2} \right) \left(\frac{1}{\omega^-} - \frac{1}{2} \right)$ the so-called magic parameter that includes the spatial error, stability, best advection and diffusion. The choice $\Lambda = \frac{1}{4}$ is suggested as a suitable value for porous media simulations. Another choice, namely $\Lambda = \frac{3}{16}$, yields the exact location of bounce-back walls in case of Poiseuille flow in a straight channel [40, 43].

2.2 Boundary Conditions

In this study, two types of boundary conditions are used for the pore-scale simulation. The first one is a no-slip wall condition and the second one is a periodic pressure forcing that is applied to drive the flow by a pressure gradient. The simplest scheme to imply no-slip boundary conditions in lattice Boltzmann is the simple bounce-back (SBB) operator. In this scheme, the wall location is represented by a staircase approximation, and the no-slip boundary is satisfied by the bounce-back phenomenon of a particle reflecting its momentum upon collision with a wall. Hence, the unknown distribution function is calculated as:

$$f_{\bar{k}}(x_{f_1}, t_{n+1}) = \tilde{f}_k(x_{f_1}, t_n). \quad (7)$$

where we take the values $\tilde{f}_k$ after collision but before streaming on the right hand side. However, the staircase approximation is not appropriate for complex geometries where more accurate results are required even for a low resolution of the boundary. Hence, the central linear interpolation (CLI) scheme which yields a higher accuracy at moderately increased computational cost is our preferred choice.

In the CLI scheme [40] three particle distribution functions are needed at two fluid nodes adjacent to the solid node, i.e.,

$$f_{\bar{k}}(x_{f_1}, t_{n+1}) = \tfrac{1-2q}{1+2q}\tilde{f}_k(x_{f_2}, t_n) - \tfrac{1-2q}{1+2q}\tilde{f}_{\bar{k}}(x_{f_1}, t_n) + \tilde{f}_k(x_{f_1}, t_n). \qquad (8)$$

while $q = |x_{f_1} - x_w|/|x_{f_1} - x_b|$ defines a normalized distance of the first fluid node to the wall. x_{f_1} and x_{f_2} are the first and second fluid neighbor cells in the direction of $\bar{k}$, respectively. We use the value $\Lambda = \tfrac{3}{16}$ for which the CLI scheme is of second order accuracy [43].

3   Large Scale Simulations

In this study, we use the WALBERLA software framework [24,44] that provides a highly optimized implementation of the TRT model that is about as fast as the SRT model. We refer to [45], where scalability of WALBERLA to more than 10^{12} lattice cells and almost 500 000 cores has been demonstrated. Compared to previous investigations, e.g., [45], we show results for the CLI scheme, which, in contrast to the SBB scheme requires data exchange with two layers of neighboring fluid cells. In WALBERLA, this situation is handled by extra ghost-layer exchanges, i.e., by communicating an extended set of distribution functions to neighboring processors. This results in an additional communication in case of massively parallel simulation runs.

To demonstrate the parallel scalability and efficiency of the WALBERLA framework in the context of a porous media simulation, we first perform a weak-scaling study. Here we use a lattice of 151^3 cells per core and embed into this grid a sphere with a diameter of 90 lattice length. The results have been obtained on the LIMA cluster at RRZE[1] which has 500 compute nodes. Each node consists of two Intel Xeon 5650 "Westmere" chips so that each node has 12 cores running at 2.66 GHz. We conduct scalability tests ranging from one node to 64 nodes. This setup results in 2.64×10^9 cells for the largest run including 768 spherical obstacles. Figure 1(a) displays the weak-scaling results using the TRT kernel. Figure 1(a) shows the mega lattice updates per second (MLUPS) for the SBB and CLI boundary schemes. The results do not only confirm that the code scales very well, but also that the MLUPS count per core compares favorably with other state of the art LBM implementations [46–48].

We point out that achieving a good scaling behavior becomes more challenging when the node performance is already high, but that a high performance on each node is a fundamental prerequisite for achieving good overall performance.

---

[1] https://www.rrze.fau.de/dienste/arbeiten-rechnen/hpc/systeme.

Thanks to both, the meticulously optimized WALBERLA kernels on each node, combined with the carefully designed communication routines, the MLUPS value per core is high and stays nearly constant while the number of cores is increased. Note that the CLI boundary conditions causes a slowdown of about 10 % in comparison to the SBB boundary condition, which is the fastest scheme. Figure 1(b) displays the percentage of the total time spent for the MPI communication, the percentage of the total time which is spent by the streaming and the collision step, and the time for the boundary handling. The slowdown of the performance while using the CLI is due to the additional time that is needed for the communication and the higher complexity of the boundary condition compared to SBB. Although, the boundary handling of the CLI scheme also takes a little bit more time than the SBB, the higher accuracy of the CLI compared to the SBB allows in complex application to use a coarser resolution of the simulation domain.

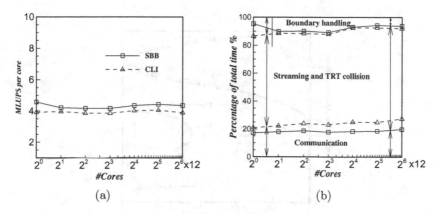

Fig. 1. Weak scaling on LIMA-Cluster using 151^3 cells per core, (a) measured MLUPS per core, (b) percentage of total time spent for MPI communication, streaming step and the TRT kernel computation, and the boundary handling step.

3.1 Pore-Scale Simulation with a Porous Medium Generated by a Particle Simulation

To construct a porous structure, we use the in-house multi-body dynamics framework PE [23]. The PE can simulate the motion of rigid bodies and their interaction by frictional collisions. Here we use this functionality to generate a random sphere packing by letting random spheres fall into the simulation domain from the top. After the spheres have come to a rest, their position is fixed and their geometry defines the solid matrix of a porous structure. The pore space is then resolved by a lattice Boltzmann grid.

The particles have different sizes and their radius is uniformly distributed in a range $[0.5D_m, 1.5D_m]$ where the parameter D_m denotes a mean diameter.

For the fluid flow simulation using the LBM, the TRT collision operator and the CLI solid boundary condition are used. This combination is fast, has second order accuracy, and shows no viscosity-dependency.

First, we test the influence of the cell size on the averaged stream-wise velocity. To do so we increase the diameter D of the spheres from 4 to 48 and keep $Re_D = \frac{U_{max}D}{\nu}$ constant. The domain has two walls at the top and bottom, and periodic boundary conditions are applied at stream-wise and span-wise directions. A constant pressure drop drives the flow, and the data are set such that $Re_D \simeq 2$. The simulation result is presented as a planar average of the stream-wise velocity in Fig. 2 while it is normalized using the maximum velocity and the height of the channel as a reference value. The results show that beyond $D = 32$ (lattice cells) a further increase of the resolution does not significantly change the results. It is worth to note that in the porous region a coarse lattice can be used and that only the transient region requires a higher resolution.

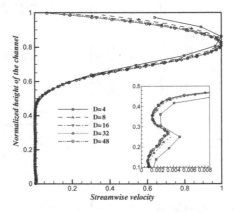

Fig. 2. Planar average stream-wise velocity for different grid sizes, $Re_D \simeq 2$.

Figure 3(b) shows the planar average stream-wise velocity for different Re numbers. To change the Re number, the viscosity and particles diameter are kept constant while the pressure gradient is changed to adjust the flow velocity. The results show that for slow flow, the velocity in the porous region is considerably higher than for fast flow. When the Re number of the flow increases, the position of the maximum velocity shifts toward to the top wall. This phenomena is due to the boundary layer effect; when the flow velocity is high in the free flow, the penetration to the porous region is less, therefore, the position of the maximum velocity changes.

In Fig. 3(b), we observe a small deviation in the velocity profile close to the bottom wall in the porous region. This is because of the high porosity close to the wall, where the spherical particles are in contact with a flat plane, see Fig. 3(a). Consequently a higher permeability can be found in this region, and the flow will accelerate because the resistance against the pressure difference is lower than in the interior of the porous medium. Therefore, to evaluate the

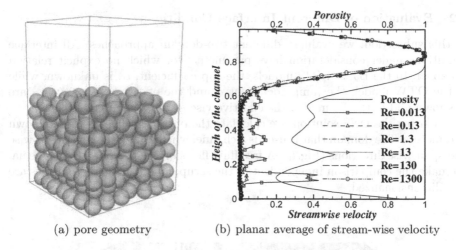

(a) pore geometry   (b) planar average of stream-wise velocity

Fig. 3. Flow over mono-sized particles for different *Re* numbers.

existing models without this effect and having a more uniform porosity in the porous region, a different set-up structure is chosen. The bottom plate of the particle simulation is placed about one particle size below the bottom wall of the fluid flow simulation. With this structure the porosity does not have the effect of placing a sphere on the wall, and therefore we create an approximately uniform permeability distribution in the porous medium.

The results of this pore-scale simulation are taken as a reference solution. Here, we use 1274 particles with radius in the range of 16–48 cells. The flow is driven by a pressure difference of 10^{-6} (in lattice units), and the simulation is run until the flow reaches the steady state. The planar average of the stream-wise velocity is depicted in Fig. 4.

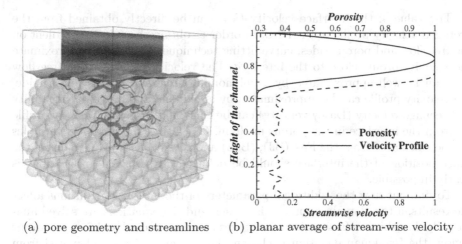

(a) pore geometry and streamlines   (b) planar average of stream-wise velocity

Fig. 4. pore-scale simulation of free flow over porous media.

3.2   Evaluation of Different Interface Conditions

In this subsection, we evaluate different two-domain approaches. All interface conditions under consideration have parameters for which no explicit relation is known. In the BJ and BJS models, the slip coefficient, α, is unknown, while in the OTW model, the jump coefficient β and the effective viscosity μ_{eff} are unknown and in the Br model, the effective viscosity μ_{eff} is unknown.

By using the DNS solution, we calculate the optimal value for the unknown parameters. The domain that is used is a channel which is periodic in stream-wise and span-wise directions (Fig. 5). A free fluid flows on the top of a porous media. To make the comparison independent of the setup, all of the flow properties are non-dimensionalized.

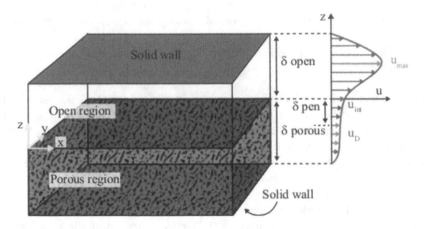

Fig. 5. Schematic of the simulation domain and averaged velocity profile in the open and porous regions.

The value of the interface velocity U_{int}, can be directly obtained from the averaged velocity profile of the DNS. In order to obtain the velocity gradient on the free flow and porous sides, curve fitting techniques are used to approximate the velocity profile close to the interface. The velocity profile on the free flow side can be well approximated by a polynomial curve and on the porous side, the velocity profile can be approximated by an exponential curve. Permeability and seepage velocity (Darcy velocity) can be calculated from the velocity profile far from the interface in the porous medium. Given this, the unknown variables can be calculated from the Eqs. (BJ), (BJS) and (OTW). However, to do so, the exact position of the interface should be defined which in real applications is nearly impossible.

To find out how the additional parameters of the interface conditions affect the results, a two-domain approach is chosen and the equations are solved analytically. For the free flow region, the Stokes equation is used and for the porous region, the Brinkman equation is chosen. The permeability is calculated from the DNS result far enough from the interface inside the porous region. In Fig. 6, we depict the planar average stream-wise velocity which is normalized based on the maximum velocity in the DNS solution.

As it can be seen in Fig. 6(a), in the Brinkman model by increasing the viscosity ratio, $J = \frac{\mu_{\text{eff}}}{\mu}$, the maximum velocity decreases and produces a discontinuity in the shear stress over the interface. In the OTW model (Fig. 6(b)), negative values of β do not influence the result significantly, however, positive values of β have a strong impact on the maximum velocity as well as on the slip velocity on the interface. Figure 6(c, d) show the results for the BJ and the BJS interface conditions. It can be observed that there is almost no difference between these two models for low Re number flows. In both these cases, the maximum velocity decreases if α increases. A small value of α results in a considerably larger maximal velocity than in the two other cases.

Quite often two-domain models result in discontinuities in the stress at the interface. Thus the a priori knowledge of the position of the interface is crucial. One possibility to fix the position of the interface is to take the location where the porosity reaches the limit value one, i.e., $y = 0.756$. However fitting of the DNS velocity profile shows that only up to $y = 0.722$, the curve is fitted well

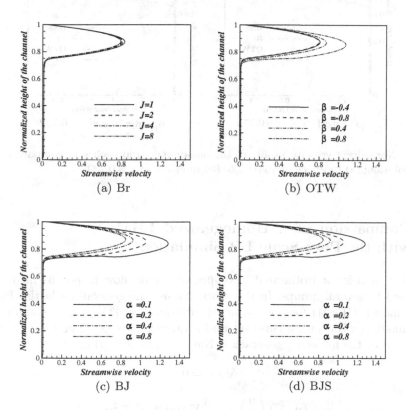

Fig. 6. Analytical solution for the velocity profile, which is normalized by the maximum velocity of the DNS solution, by different interface models.

by an exponential function. More precisely, $u(h) = 0.48423 \cdot \exp(0.31195\,h) - 0.48236 \cdot \exp(0.3131\,h)$ yields a root mean squared error of $5.736 \cdot 10^{-6}$. The pure fluid flow velocity profile is fitted to a 2nd order polynomial resulting in $u(h) = (1.9593e{-}3) + (2.78421e{-}4)h - (4.48066e{-}6)h^2$ with a root mean squared error of $9.5815 \cdot 10^{-6}$. This observation motivates an alternative choice of the interface position where the corresponding governing equations will be fulfilled. Calculating the slip coefficient and the jump coefficient for these two positions, we find for $y = 0.756$, $\alpha = 0.3163$, $\beta = -2.8397$ and for $y = 0.722$, $\alpha = 0.31645$ and $\beta = -2.8397$. However, as it can be seen in Fig. 7, even with the parameters which are extracted from the DNS results, the considered two-domain approaches cannot represent accurately the DNS solution. Comparing Fig. 7(a, b) shows that the two-domain approaches depend strongly on the interface position and more sophisticated criteria for defining the interface location are required to obtain better matching results.

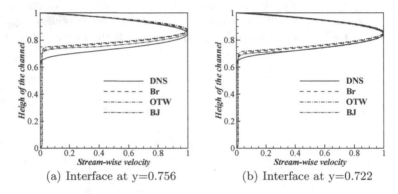

(a) Interface at y=0.756   (b) Interface at y=0.722

Fig. 7. Normalized velocity profile of the one-domain approaches in compare to the DNS solution; (a) interface at $y = 0.756$, (b) interface at $y = 0.722$.

4   Comparsion of a Homogenized LBM with the Pore-Scale LB Simulation

Different models for isothermal incompressible fluid flow in porous media are proposed by several groups. In this work, we use the generalized lattice Boltzmann model (GLBM) for porous media introduced in [35], which is applicable for a medium with both a constant and a variable porosity. The model can be expressed by the following generalized Navier-Stokes equation:

$$\nabla \cdot \mathbf{u} = 0 \tag{9}$$

$$\frac{\partial \mathbf{u}}{\partial t} + (\mathbf{u} \cdot \nabla)\left(\frac{\mathbf{u}}{\epsilon}\right) = -\frac{1}{\rho}\nabla\left(\epsilon p\right) + \nu_{\text{eff}}\nabla^2\mathbf{u} + \mathbf{F}, \tag{10}$$

where ρ is the fluid density, $\mathbf{u}$ and p are the volume-averaged velocity and pressure, respectively, ν_{eff} is the effective viscosity, and ϵ is the porosity. The total body force $\mathbf{F}$ caused by the presence of a porous medium and other external force fields is given by

$$\mathbf{F} = -\frac{\epsilon\nu}{K}\mathbf{u} - \frac{\epsilon c_F}{\sqrt{K}}|\mathbf{u}|\mathbf{u} + \epsilon\mathbf{G}, \tag{11}$$

where ν is the shear viscosity of the fluid that is not necessarily the same as ν_{eff}, $\mathbf{G}$ is the body force induced by an external force, c_F is the Forchheimer coefficient that depends on the porous structure, and K is the permeability of the porous media. The first and the second terms on the right hand side of Eq. (11) are the linear Darcy and non-linear Forchheimer drags due to the porous medium, respectively. The quadratic nature of the non-linear resistance makes it negligible for low-speed flows, but is more noteworthy in hindering the fluid motion for high-speed flows, i.e., high Re number and high Da number flows.

The GLBM considers Eq. (11) as a source term in Eq. (2) and also modifies the equilibrium distribution function (Eq. (5)) based on the porosity. The detailed formulation can be found in [35].

Firstly to validate the generalized model for flow over a porous medium, we choose a simple Couette flow. The lower-half of the channel of width H is filled with a porous medium with a porosity of ϵ, the stream-wise and spanwise boundaries are periodic, and the top wall of the channel is moving with a constant velocity of u_0. Then, the steady state velocity in this channel satisfies

$$\nu_{\text{eff}}\nabla^2\mathbf{u} - \frac{\epsilon\nu}{K}\mathbf{u} - \frac{\epsilon c_F}{\sqrt{K}}|\mathbf{u}|\mathbf{u} + \epsilon\mathbf{G} = 0, \tag{12}$$

while the walls of the channel are modeled by a no-slip condition.

Figure 8 shows the velocity profile for the Couette flow with different viscosity ratios J $(= \mu_e/\mu)$ and compared to a semi-analytical solution for $Re = 0.1$ and

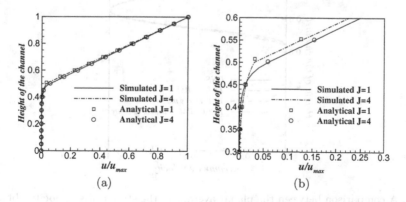

(a)   (b)

Fig. 8. Velocity profile of the Couette flow for different viscosity ratios $J = \mu_e/\mu$, in comparison with the approximate analytical solution of Eq. (13), (a) global system; (b) zoom into the region near the interface

$Da = 0.00012$. In the Stokes regime for a low Da number, [18] reported that the velocity profile in the free flow is linear and exponentially decaying in the porous region. More precisely the semi-analytic solution can be written as:

$$u_x(y) = \begin{cases} rKa + \epsilon a\,(y - H/2) & H/2 \leq y \leq H \\ rKae^{r(y-H/2)} & 0 \leq y \leq H/2 \end{cases} \tag{13}$$

where

$$a = \frac{2u_0}{2rK + \epsilon H}, \quad r = \frac{\sqrt{\nu\epsilon}}{\sqrt{\nu_{\text{eff}}k}}, \tag{14}$$

and u_0 is the lid's velocity. The simulation result shows excellent agreement with the analytical solution for both viscosity ratios.

Secondly, we apply the generalized model to a problem with no sharp interface and a significant porosity change close to the interface. We use the planar average of the porosity as it is obtained in the DNS, therefore, there is no need to explicitly set the interface position. Since the flow is within the Stokes regime, the Forchheimer term in Eq. (11) is neglected.

Figure 9 shows the results of the planar average stream-wise velocity for the DNS solution and the GLBM. Although the porosity, permeability, fluid properties and driving forces are the same, the standard GLBM homogenized model over-predicts the velocity in the transition zone. The dashed line shows the homogenized model that only takes the Darcy force into account. These two mentioned homogenized models use a viscosity in the porous region which is equal to the free flow region. We propose to use the GLBM homogenized model but with a viscosity in the porous region depending on the porosity by $\mu_{\text{eff}} = \mu/\epsilon$. As we can observe in the porous region, the latter model can perfectly predict the DNS result.

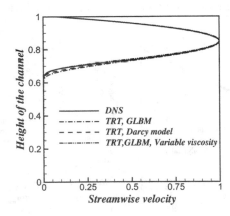

Fig. 9. A comparison between the planar average of the stream-wise velocity obtained by DNS and the homogenized model, $Re_D \simeq 2$.

5   Conclusion

We presented three different approaches to simulate the interaction of free flow with porous media flow, namely, direct pore-scale simulations, as well as homogenized single-domain and two-domains approaches. The lattice Boltzmann method is employed both, for obtaining the pore-scale reference solution, and for solving the computationally more appealing homogenized problems.

For the two-domain approaches, four different interface conditions for dealing with the physical transport through a sharp interface have been evaluated. Our comparison yields that the two-domain techniques are quite sensitive to the interface position. To further investigate this effect, we examined two definitions for the interface position, i.e., the exact and the apparent position assumptions. However, as our results indicate, both approaches fall short with respect to accuracy in the vicinity of the interface if the exact interface geometry is unknown. As an alternative approach we consider a homogenized one-domain model that is based on the idea of a smooth transition zone between the free flow and porous media models. A simple porosity-dependent rescaling of the viscosity allows us to accurately reproduce the results obtained by averaging the pore-scale solution.

In future work we aim to investigate the combination of both approaches to allow for the treatment of more general situations in a two-scale fashion. Since the discussed lattice Boltzmann schemes are suitable for REV-scale computations, and are also highly scalable for pore-scale simulations, they lend themselves well for leveraging the power of massively parallel computing architectures.

Acknowledgement. Financial support from the German Research Foundation (DFG, Project WO 671/11-1) and also the International Graduate School of Science and Engineering (IGSSE) of the Technische Universität München for research training group 6.03 are gratefully acknowledged. Our special thank goes to Regina Ammer for fruitful discussions and the WALBERLA primary authors Florian Schornbaum, Christian Godenschwager and Martin Bauer for their essential help with implementing the code.

References

1. Helmig, R.: Multiphase Flow and Transport Processes in the Subsurface: A Contribution to the Modeling of Hydrosystems. Springer, Heidelberg (2011)
2. Alazmi, B., Vafai, K.: Analysis of fluid flow and heat transfer interfacial conditions between a porous medium and a fluid layer. Int. J. Heat Mass Transf. **44**, 1735–1749 (2001)
3. Nield, D., Kuznetsov, A.: The effect of a transition layer between a fluid and a porous medium: shear flow in a channel. Transp. Porous Media **78**, 477–487 (2009)
4. Le Bars, M., Worster, M.G.: Interfacial conditions between a pure fluid and a porous medium: implications for binary alloy solidification. J. Fluid Mech. **550**, 149–173 (2006)
5. Goyeau, B., Lhuillier, D., Gobin, D., et al.: Momentum transport at a fluid-porous interface. Int. J. Heat Mass Transf. **46**, 4071–4081 (2003)

6. Chandesris, M., Jamet, D.: Jump conditions and surface-excess quantities at a fluid/porous interface: a multi-scale approach. Transp. Porous Media **78**, 419–438 (2009)
7. Goharzadeh, A., Khalili, A., Jørgensen, B.B.: Transition layer thickness at a fluid-porous interface. Phys. Fluids **17**, 057102 (2005)
8. Ghisalberti, M.: The three-dimensionality of obstructed shear flows. Environ. Fluid Mech. **10**, 329–343 (2010)
9. Morad, M., Khalili, A.: Transition layer thickness in a fluid-porous medium of multi-sized spherical beads. Exp. Fluids **46**, 323–330 (2009)
10. Pokrajac, D., Manes, C.: Velocity measurements of a free-surface turbulent flow penetrating a porous medium composed of uniform-size spheres. Transp. Porous Media **78**, 367–383 (2009)
11. Beavers, G.S., Joseph, D.D.: Boundary conditions at a naturally permeable wall. J. Fluid Mech. **30**, 197–207 (1967)
12. Nield, D., Bejan, A.: Convection in Porous Media. Springer, New York (2006)
13. Duman, T., Shavit, U.: An apparent interface location as a tool to solve the porous interface flow problem. Transp. Porous Media **78**, 509–524 (2009)
14. Baber, K., Mosthaf, K., Flemisch, B., Helmig, R., Müthing, S., Wohlmuth, B.: Numerical scheme for coupling two-phase compositional porous-media flow and one-phase compositional free flow. IMA J. Appl. Math. **6**, 887–909 (2012)
15. Mosthaf, K., Baber, K., Flemisch, B., Helmig, R., Leijnse, A., Rybak, I., Wohlmuth, B.: A new coupling concept for two-phase compositional porous media and single-phase compositional free flow. Water Resour. Res. **47**, 1–19 (2011)
16. Saffman, P.: On the boundary condition at the surface of a porous medium. Stud. Appl. Math. **50**, 93–101 (1971)
17. Ochoa-Tapia, J., Whitaker, S.: Momentum transfer at the boundary between a porous medium and a homogeneous fluid – II. Comparison with experiment. Int. J. Heat Mass Transf. **38**, 2647–2655 (1995)
18. Martys, N., Bentz, D.P., Garboczi, E.J.: Computer simulation study of the effective viscosity in Brinkman's equation. Phys. Fluids **6**, 1434–1439 (1994)
19. Lundgren, T.S.: Slow flow through stationary random beds and suspensions of spheres. J. Fluid Mech. **51**, 273–299 (1972)
20. Zhang, Q., Prosperetti, A.: Pressure-driven flow in a two-dimensional channel with porous walls. J. Fluid Mech. **631**, 1–21 (2009)
21. Nabovati, A., Amon, C.: Hydrodynamic boundary condition at open-porous interface: a pore-level lattice Boltzmann study. Transp. Porous Media **96**, 83–95 (2013)
22. Liu, Q., Prosperetti, A.: Pressure-driven flow in a channel with porous walls. J. Fluid Mech. **679**, 77–100 (2011)
23. Preclik, T., Rüde, U.: Ultrascale simulations of non-smooth granular dynamics. Comput. Part. Mech. 1–24 (2015)
24. Feichtinger, C., Götz, J., Donath, S., Iglberger, K., Rüde, U.: Walberla: exploiting massively parallel systems for lattice Boltzmann simulations. In: Trobec, R., Vajteršic, M., Zinterhof, P. (eds.) Parallel Computing, pp. 241–260. Springer, London (2009)
25. Rong, L.W., Dong, K.J., Yu, A.B.: Lattice-Boltzmann simulation of fluid flow through packed beds of spheres: effect of particle size distribution. Chem. Eng. Sci. **116**, 508–523 (2014)
26. Beetstra, R., van der Hoef, M.A., Kuipers, J.A.M.: TA Lattice-Boltzmann simulation study of the drag coefficient of clusters of spheres. Comput. Fluids **35**, 966–970 (2006)

27. Succi, S., Foti, E., Higuera, F.: Three-dimensional flows in complex geometries with the lattice Boltzmann method. EPL (Europhys. Lett.) **10**, 433 (1989)
28. Singh, M., Mohanty, K.: Permeability of spatially correlated porous media. Chem. Eng. Sci. **55**, 5393–5403 (2000)
29. Bernsdorf, J., Brenner, G., Durst, F.: Numerical analysis of the pressure drop in porous media flow with lattice Boltzmann (BGK) automata. Comput. Phys. Commun. **129**, 247–255 (2000)
30. Kim, J., Lee, J., Lee, K.C.: Nonlinear correction to Darcy's law for a flow through periodic arrays of elliptic cylinders. Physica A: Stat. Mech. Appl. **293**, 13–20 (2001)
31. Spaid, M.A.A., Phelan, F.R.: Lattice Boltzmann methods for modeling microscale flow in fibrous porous media. Phys. Fluids **9**, 2468–2474 (1997)
32. Freed, D.M.: Lattice-Boltzmann method for macroscopic porous media modeling. Int. J. Mod. Phys. C **09**, 1491–1503 (1998)
33. Martys, N.S.: Improved approximation of the Brinkman equation using a lattice Boltzmann method. Phys. Fluids **13**, 1807–1810 (2001)
34. Nithiarasu, P., Seetharamu, K., Sundararajan, T.: Natural convective heat transfer in a fluid saturated variable porosity medium. Int. J. Heat Mass Transf. **40**, 3955–3967 (1997)
35. Guo, Z., Zhao, T.: Lattice Boltzmann model for incompressible flows through porous media. Phys. Rev. E **66**, 036304 (2002)
36. Bhatnagar, P.L., Gross, E.P., Krook, M.: A model for collision processes in gases. I. Small amplitude processes in charged and neutral one-component systems. Phys. Rev. **94**, 511–525 (1954)
37. Pan, C., Luo, L.S., Miller, C.T.: An evaluation of lattice Boltzmann schemes for porous medium flow simulation. Comput. Fluids **35**, 898–909 (2006)
38. Bogner, S., Mohanty, S., Rüde, U.: Drag correlation for dilute and moderately dense fluid-particle systems using the lattice Boltzmann method. Int. J. Multiph. Flow **68**, 71–79 (2015)
39. Ginzburg, I.: Lattice Boltzmann modeling with discontinuous collision components: hydrodynamic and advection diffusion equations. J. Stat. Phys. **126**, 157–206 (2007)
40. Ginzburg, I., Verhaeghe, F., d'Humieres, D.: Two-relaxation-time lattice Boltzmann scheme: about parametrization, velocity, pressure and mixed boundary conditions. Commun. Comput. Phys. **3**, 427–478 (2008)
41. Ginzburg, I., Verhaeghe, F., d'Humières, D.: Study of simple hydrodynamic solutions with the two-relaxation-times lattice-Boltzmann scheme. Commun. Comput. Phys. **3**, 519–581 (2008)
42. He, X., Luo, L.S.: Lattice Boltzmann model for the incompressible Navier-Stokes equation. J. Stat. Phys. **88**, 927–944 (1997)
43. Khirevich, S., Ginzburg, I., Tallarek, U.: Coarse- and fine-grid numerical behavior of MRT/TRT lattice Boltzmann schemes in regular and random sphere packings. J. Comput. Phys. **281**, 708–742 (2015)
44. Feichtinger, C., Donath, S., Köstler, H., Götz, J., Rüde, U.: WaLBerla: HPC software design for computational engineering simulations. J. Comput. Sci. **2**, 105–112 (2011)
45. Godenschwager, C., Schornbaum, F., Bauer, M., Köstler, H., Rüde, U.: A framework for hybrid parallel flow simulations with a trillion cells in complex geometries. In: Proceedings of SC13: International Conference for High Performance Computing, Networking, Storage and Analysis, SC 2013, NY, USA, pp. 35:1–35:12. ACM, New York (2013)

46. Peters, A., Melchionna, S., Kaxiras, E., Lätt, J., Sircar, J., Bernaschi, M., Bison, M., Succi, S.: Multiscale simulation of cardiovascular flows on the IBM Blue-gene/P: full heart-circulation system at red-blood cell resolution. In: Proceedings of the 2010 ACM/IEEE International Conference for High Performance Computing, Networking, Storage and Analysis, pp. 1–10. IEEE Computer Society (2010)
47. Schönherr, M., Kucher, K., Geier, M., Stiebler, M., Freudiger, S., Krafczyk, M.: Multi-thread implementations of the lattice Boltzmann method on non-uniform grids for CPUs and GPUs. Comput. Math. Appl. **61**, 3730–3743 (2011)
48. Robertsen, F., Westerholm, J., Mattila, K.: Lattice Boltzmann simulations at petascale on multi-GPU systems with asynchronous data transfer and strictly enforced memory read alignment. In: 23rd Euromicro International Conference on Parallel, Distributed and Network-Based Processing (PDP), pp. 604–609 (2015)

Chrono: An Open Source Multi-physics Dynamics Engine

Alessandro Tasora[1], Radu Serban[2], Hammad Mazhar[2], Arman Pazouki[2],
Daniel Melanz[2], Jonathan Fleischmann[2], Michael Taylor[2],
Hiroyuki Sugiyama[3], and Dan Negrut[2(✉)]

[1] University of Parma, Parma, Italy
[2] University of Wisconsin–Madison, Madison, WI, USA
`negrut@wisc.edu`
[3] University of Iowa, Iowa City, IA, USA

Abstract. We provide an overview of a multi-physics dynamics engine
called Chrono. Its forte is the handling of complex and large dynamic systems containing millions of rigid bodies that interact through frictional
contact. Chrono has been recently augmented to support the modeling
of fluid-solid interaction (FSI) problems and linear and nonlinear finite
element analysis (FEA). We discuss Chrono's software layout/design and
outline some of the modeling and numerical solution techniques at the
cornerstone of this dynamics engine. We briefly report on some validation studies that gauge the predictive attribute of the software solution.
Chrono is released as open source under a permissive BSD3 license and
available for download on GitHub.

Keywords: Multi-physics modeling and simulation · Rigid and flexible
multi-body dynamics · Friction and contact · Fluid solid interaction ·
Granular dynamics · Vehicle dynamics · Parallel computing

1 Overview of Chrono and its Software Infrastructure

Chrono is an open source software infrastructure used to investigate the time
evolution of systems governed by very large sets of differential-algebraic equations and/or ordinary differential equations and/or partial differential equations
[48,70]. Chrono can currently be used to simulate (*i*) the dynamics of large
systems of connected bodies governed by differential-algebraic equations; (*ii*)
controls and other first-order dynamic systems governed by ordinary differential equations; (*iii*) fluid–solid interaction problems governed, in part, by the
Navier-Stokes equations; and (*iv*) the dynamics of deformable bodies governed
by partial differential equations. Chrono can handle multi-physics problems in
which the solution calls for the mixing of (*i*) through (*iv*).

This dynamics simulation engine rests on five foundation components that
provide the following basic functionality: equation formulation, equation solution, collision detection and proximity computation, support for parallel computing, and pre/post-processing, see Fig. 1. The first foundation component, called

© Springer International Publishing Switzerland 2016
T. Kozubek et al. (Eds.): HPCSE 2015, LNCS 9611, pp. 19–49, 2016.
DOI: 10.1007/978-3-319-40361-8_2

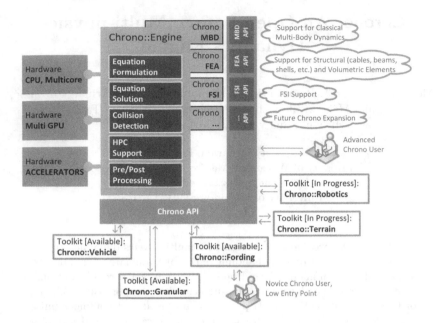

Fig. 1. An abstraction of the Chrono architecture. Chrono provides support for three application areas, namely multibody dynamics (MBD), FEA, and FSI, building on a five component foundation that handles the formulation of the equation of motion, numerical solution, collision detection, parallel and high-performance computing support, and pre/post processing tasks. An API (in blue) allows expert users to interact with Chrono. Toolkits such as Chrono::Vehicle, Chrono::Granular, etc. provide a low-entry point for users who need domain-specific Chrono support (Color figure online).

Equation Formulation, supports general-purpose modeling for large systems of rigid and flexible bodies and for basic FSI problems. The second component, called *Equation Solution*, provides the algorithmic support needed to numerically solve the resulting equations of motion. Proximity computation support, essential for collision detection and computation of short range interaction forces, is provided by the third foundation component. The fourth component enables the partitioning and farming out of very large dynamics problems for parallel execution on supercomputer architectures using the Message Passing Interface (MPI) paradigm [58]. The fifth component provides pre- and post-processing support.

Chrono is almost entirely written in C++. It is compiled into a library subsequently used by third party applications. A user can invoke functions implemented in Chrono via an Application Programming Interface (API) that comes in two options: C++ and Python. Chrono runs on Windows, Linux, and Mac OSX and is designed to leverage parallel computing. Depending on the solution mode in which it operates, it can rely on parallel computing on Graphics Processing Unit (GPU) cards using CUDA [61], multi-core computing using OpenMP [62], and multi-node parallel computing using MPI [57].

Chrono has been validated against experimental data, analytical results, and commercial software. Two correlation studies against MSC.ADAMS are summarized in [75,88]. The gradient-deficient beams and plates in Chrono::FEA have been validated against the commercial code ABAQUS [51]. The frictional-contact solution in Chrono has been validated against experimental data for angle of repose simulations [47], rate of flow [34,52], impact tests [49,73], and shear tests [24]. Finally, the fledgling Chrono::FSI module has been validated against analytical results and experimental data [65].

Chrono has been used for tracked and wheeled vehicle dynamics, in robotics (optimization and design of parallel kinematics machines), in the field of seismic engineering (simulation of earthquake effects on ancient buildings), in the field of waste processing (granular flows in separation machines), in additive manufacturing and 3D printing (see Fig. 2a and b), to simulate new types of escapements in mechanical clocks, and to characterize ice sheet dynamics for the oil industry. Approximately 150 movies that illustrate simulations run in Chrono are available at [78,79].

In terms of user support, the API documentation for the main Chrono modules is generated from their annotated C++ sources using Doxygen [20]. All Chrono software is configured and built using CMake [38] for a robust cross-platform build experience under Linux, Mac OSX, and Windows. Unit testing relies on the CTest suite of tools [38] to automate configuring, building, and executing the tests. Chrono operates under a continuous integration mode based on Buildbot [12]. Results and output from all builds, tests, and benchmarks are recorded to an external database accessible to all developers.

2   Rigid Body Dynamics Support in Chrono

The dynamics of articulated systems composed of rigid and flexible bodies is characterized by a system of index 3 Differential Algebraic Equations (DAEs) [31,33,74] shown using the terms in black font in Eqs. (1a)–(1c):

$$\dot{\mathbf{q}} = \mathbf{L}(\mathbf{q})\mathbf{v} \tag{1a}$$

$$\mathbf{M}(\mathbf{q})\dot{\mathbf{v}} = \mathbf{f}(t, \mathbf{q}, \mathbf{v}) - \mathbf{g}_{\mathbf{q}}^{T}(\mathbf{q}, t)\hat{\lambda} + \sum_{i \in \mathcal{A}(\mathbf{q}, \delta)} \underbrace{(\hat{\gamma}_{i,n}\, \mathbf{D}_{i,n} + \hat{\gamma}_{i,u}\, \mathbf{D}_{i,u} + \hat{\gamma}_{i,w}\, \mathbf{D}_{i,w})}_{i^{th}\text{frictional contact force}} \tag{1b}$$

$$\mathbf{0} = \mathbf{g}(\mathbf{q}, t) \tag{1c}$$

$$i \in \mathcal{A}(\mathbf{q}(t)): \begin{cases} 0 \leq \Phi_i(\mathbf{q}) \perp \hat{\gamma}_{i,n} \geq 0 \\ (\hat{\gamma}_{i,u}, \hat{\gamma}_{i,w}) = \underset{\sqrt{\bar{\gamma}_{i,u}^2 + \bar{\gamma}_{i,w}^2} \leq \mu_i \hat{\gamma}_{i,n}}{\operatorname{argmin}} \mathbf{v}^T \cdot (\bar{\gamma}_{i,u}\, \mathbf{D}_{i,u} + \bar{\gamma}_{i,w}\, \mathbf{D}_{i,w}) \,. \end{cases} \tag{1d}$$

The differential equations in (1a) relate the time derivative of the generalized positions $\mathbf{q}$ and velocities $\mathbf{v}$ through a linear transformation defined by $\mathbf{L}(\mathbf{q})$. The force balance equation in (1b) ties the inertial forces to the applied and constraint forces, $\mathbf{f}(t, \mathbf{q}, \mathbf{v})$ and $-\mathbf{g}_{\mathbf{q}}^{T}(\mathbf{q}, t)\hat{\lambda}$, respectively. The latter are imposed by bilateral constraints that restrict the relative motion of the rigid or flexible bodies present in the system. These bilateral constraints, which lead to Eq. (1c), can be augmented by unilateral constraints associated with contact/impact phenomena. To that end, the concept of equations of motion is extended to employ differential inclusions [23]. The simplest example is that of a body that interacts with the ground through friction and contact, when the equations of motion become an inclusion $\mathbf{M\ddot{q}} - \mathbf{f} \in \mathbf{F}(\mathbf{q}, t)$, where $\mathbf{M}$ is the inertia matrix, $\mathbf{\ddot{q}}$ is the body acceleration, $\mathbf{f}$ is the external force, and $\mathbf{F}(\mathbf{q}, t)$ is a set-valued function. The inclusion states that the frictional contact force lies somewhere inside the friction cone, with a value yet to be determined and controlled by the stick/slip state of the interaction between body and ground. In MBD the differential inclusion can be posed as a differential variational inequality problem [81], which brings along the red-font terms in Eq. (1) [2,5,30,37,82,89]. Specifically, the unilateral constraints define a set of contact complementarity conditions $0 \leq \Phi_i(\mathbf{q}) \perp \widehat{\gamma}_{i,n} \geq 0$, which make a simple point: for a potential contact i in the active set, $i \in \mathcal{A}(\mathbf{q}(t))$, either the gap Φ_i between two geometries is zero and consequently the normal contact force $\widehat{\gamma}_{i,n}$ is greater than zero, or vice versa. The last equation poses an optimization problem whose first order Karush-Kuhn-Tucker optimality conditions are equivalent to the Coulomb dry friction model [81]. The frictional contact force associated with contact i leads to a set of generalized forces, shown in red in Eq. (1b), which are obtained using the projectors $\mathbf{D}_{i,n}$, $\mathbf{D}_{i,u}$, and $\mathbf{D}_{i,w}$ [6].

The modeling methodology outlined above has been used in Chrono to analyze the dynamics of large multibody systems and granular material in a so called Discrete Element Method (DEM) framework. Since the methodology uses complementarity (C) conditions to enforce non-penetration of the discrete elements that come in mutual contact, this method is called DEM-C. This differentiates it from DEM-P, a penalty (P) based methodology that is also implemented in Chrono and which accounts for the partial deformation of the bodies in mutual contact. Chrono, in its DEM-C embodiment that draws on Eq. (1), is shown at work in conjunction with two additive manufacturing processes in Fig. 2. Several frames of the Selective Laser Sintering (SLS) layering process simulations for various translational speeds are juxtaposed in Fig. 2(a). The model consists of 1300000 rigid spheres with an average diameter of $55\,\mu\text{m}$ and a density of $930\,\text{kg/m}^3$ [47]. The radius of the powder particles was randomly distributed using a normal distribution. A roller with a diameter of .0762 m travels at various longitudinal speeds and rotates at a rate of 3.33 rad/s. An approximate equation tally is as follows: close to eight million from (1a), the same count for (1b), and almost no equations for (1c). In (1d), there are approximately 7.8 million contact complementarity conditions for the normal force and, because the problem accounts for normal, sliding, rolling and spinning of the bodies, there are six Lagrange multipliers for each contact. The optimization problem that provided the frictional contact forces for this simulation was posed in approximately 46.8 million variables and solved in Chrono using a methodology proposed in [34,49]. The handling of the rolling and spinning friction is described in [86]. The image in Fig. 2(b) shows a frame of a folding simulation that is a key step in predicting where each component of a dress is located inside the 3D printing volume. The position and orientation of each element is passed to the 3D printer which then prints the entire dress. The chain-mail dress is made up of 40760 rigid rings and 55 clasps collapsing into a $10 \times 10 \times 10$ in^3 printing volume [50].

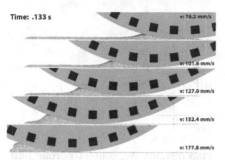

(a) Chrono simulation of SLS layering process.

(b) Chrono simulation of dress folding.

Fig. 2. Left – frames of the SLS layering process simulations for various translational speeds. Right – a frame of a folding simulation used in the 3D printing of a dress.

When using the DEM-P approach, or soft-body approach, Chrono regards the contacting bodies are "soft" in the sense that they are allowed to "overlap" or experience local deformation before a corrective contact force is applied at the point of contact. Once such an overlap δ_n is detected, by any one of a number of contact algorithms, contact force vectors $\mathbf{F}_n$ and $\mathbf{F}_t$ normal and tangential to the contact plane at the point of contact are calculated using various constitutive laws [39,40,43] based on the local body deformation at the point of contact. In the contact-normal direction, $\mathbf{n}$, this local body deformation is defined as the penetration (overlap) of the two quasi-rigid bodies, $\mathbf{u}_n = \delta_n \mathbf{n}$. In the contact-tangential direction, the deformation is defined as a vector $\mathbf{u}_t$ that tracks the total tangential displacement of the initial contact points on the two quasi-rigid bodies, projected onto the current contact plane, as shown in Fig. 3.

An example of a DEM-P contact constitutive law, a slightly modified form of which is used in Chrono, is the following viscoelastic model based on either Hookean or Hertzian contact theory:

$$\begin{aligned}
\mathbf{F}_n &= f(\bar{R}, \delta_n)\left(k_n \mathbf{u}_n - \gamma_n \bar{m} \mathbf{v}_n\right) \\
\mathbf{F}_t &= f(\bar{R}, \delta_n)\left(-k_t \mathbf{u}_t - \gamma_t \bar{m} \mathbf{v}_t\right),
\end{aligned} \tag{2}$$

where $\mathbf{u} = \mathbf{u}_n + \mathbf{u}_t$ is the overlap or local contact displacement of two interacting bodies. The quantities $\bar{m} = m_i m_j / (m_i + m_j)$ and $\bar{R} = R_i R_j / (R_i + R_j)$ represent the effective mass and effective radius of curvature, respectively, for contacting bodies with masses m_i and m_j and contact radii of curvature R_i and R_j. The vectors $\mathbf{v}_n$ and $\mathbf{v}_t$ are the normal and tangential components of the relative

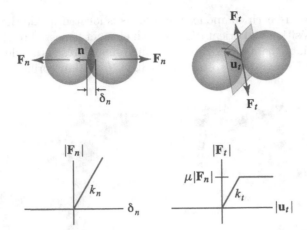

Fig. 3. DEM-P contact model, with normal overlap distance δ_n, contact-normal unit vector **n**, and tangential displacement vector $\mathbf{u}_t$ in the plane of contact (top), and a Hookean-linear contact force-displacement law with constant Coulomb sliding friction (bottom).

velocity at the contact point. For Hookean contact, $f(\bar{R}, \delta_n) = 1$ in Eq. (2); for Hertzian contact, one can let $f(\bar{R}, \delta_n) = \sqrt{\bar{R}\delta_n}$ [43, 76, 93]. The normal and tangential stiffness and damping coefficients k_n, k_t, γ_n, and γ_t are obtained, through various constitutive laws derived from contact mechanics, from physically measurable properties for the materials of the contacting bodies, such as Young's modulus, Poisson's ratio, the coefficient of restitution, etc. Detailed descriptions of the DEM-P contact models implemented in Chrono, as well as alternative DEM-P contact models, are provided in [24].

The component of the contact displacement vector **u** in the contact-normal direction, $\mathbf{u}_n = \delta_n \mathbf{n}$, is obtained directly from the contact detection algorithm, which provides the magnitude of the "inter-penetration" δ_n between the bodies. The tangential contact displacement vector $\mathbf{u}_t$ is formulated as

$$\mathbf{u}_t = \int_{t_0}^{t} \mathbf{v}_t dt - \left(\mathbf{n} \cdot \int_{t_0}^{t} \mathbf{v}_t \, dt \right) \mathbf{n}, \tag{3}$$

where t is the current time and t_0 is the time at the initiation of contact [28]. For the true tangential contact displacement history model, the vector $\mathbf{u}_t$ must be stored and updated at each time step for each contact point on a given pair of contacting bodies from the time that contact is initiated until that contact is broken. On the importance of tangential contact displacement history in DEM-P contact models, see [25].

To enforce the Coulomb friction law, if $|\mathbf{F}_t| > \mu|\mathbf{F}_n|$ at any given time step, then before the contributions of the contact forces are added to the resultant force and torque on the body, the (stored) value of $|\mathbf{u}_t|$ is scaled so that $|\mathbf{F}_t| = \mu|\mathbf{F}_n|$, where μ is the Coulomb (static and sliding) friction coefficient. For example, if $f(x) = 1$ in Eq. (2), then

$$k_t|\mathbf{u}_t| > \mu|\mathbf{F}_n| \quad \Rightarrow \quad \mathbf{u}_t \leftarrow \mathbf{u}_t \frac{\mu|\mathbf{F}_n|}{k_t|\mathbf{u}_t|} . \tag{4}$$

Once the contact forces $\mathbf{F}_n$ and $\mathbf{F}_t$ are computed for each contact and their contributions are summed to obtain a resultant force and torque on each body in the system, the time evolution of each body in the system is obtained by integrating the Newton-Euler equations of motion, subject to the Courant-Friedrichs-Lewy (CFL) stability condition, which limits [63] the integration time step-size to $h < h_{\text{crit}} \sim \sqrt{m_{\text{min}}/k_{\text{max}}}$.

For multibody dynamics without frictional contact, or with frictional contact modeled using a penalty approach, Chrono implements index 3 DAE solutions [35,59]. For handling frictional contact within the differential variational inequality framework, Chrono implements a variant of the Nesterov algorithm [49]. Handling of cohesion is based on the an approach described in [87].

3   Flexible Body Dynamics Support in **Chrono**

Chrono implements flexible body dynamics support for problems in which bodies are expected to sustain large deformations that take place while the body might experience large translational and/or rotational accelerations. The current implementation does not support the so-called floating-frame-of-reference, modal, approach, in which the small deformation of the body is superimposed on top of a large reference motion, see for instance [72]. For large deformations, Chrono currently resorts to the Absolute Nodal Coordinate Formulation (ANCF) for structural elements such as beams, plates, and shells and a corotational (CR) approach for both structural and volumetric elements.

3.1   Nonlinear Finite Element Analysis via the Absolute Nodal Coordinate Formulation

ANCF has proven to be successful in solving various challenging engineering problems of complex multibody systems including cables, belt drives, rotor blades, leaf springs, tires and many others [27,74,83,91]. In this finite element formulation, the large rotation and deformation of the element are parameterized by the global position vector gradients, and no rotational nodal coordinates such as Euler angles are utilized. This parameterization leads to a constant mass matrix for fully nonlinear dynamic problems while ensuring the exact modeling of rigid body reference motion [74]. Using the polar decomposition theorem, the displacement gradient tensor can be decomposed into the orthogonal rotation matrix and the stretch tensor that describes the most general six deformation modes. That is, use of the position vector gradient coordinates allows for describing the rotation and deformation within the element, thereby circumventing the complex nonlinear coupling of the rotation and deformation coordinates that appears in the inertia terms of flexible body models using rotational parameterization. The constant mass matrix of large deformable bodies not only leads to efficient solutions in nonlinear dynamics simulation, but also allows for the use of the non-incremental solution procedures utilized in general multibody dynamics computer algorithms. Using these important features and general motion description employed in ANCF, the structural beam and plate/shell elements can be implemented in the general multibody dynamics computer algorithms of Chrono without resorting to ad hoc co-simulation procedures. In what follows, the beam and plate/shell elements implemented in Chrono are summarized and important features of these elements are highlighted.

Thin Beam Element. The beam element implemented in Chrono is suited for modeling thin beam structures such as cables and belt drives, in which the transverse shearing effect of the beam cross section is considered negligible. The global position vector of the two-node ANCF Euler-Bernoulli beam element i on the centerline is defined by [26]

$$\mathbf{r}^i = \mathbf{S}^i(x^i)\mathbf{e}^i , \tag{5}$$

where $\mathbf{S}^i$ is the element shape function matrix obtained by the cubic Hermite polynomial [26], x^i is the element axial coordinate, and $\mathbf{e}^i$ is the nodal coordinate vector of element i. Node j of element i has the global position vector $\mathbf{r}^{ij}$ and global position vector gradient vector $\partial \mathbf{r}^{ij}/\partial x^i$ that is tangent to the beam centerline as follows [26]:

$$\mathbf{e}^{ij} = \left[(\mathbf{r}^{ij})^T \quad (\frac{\partial \mathbf{r}^{ij}}{\partial x^i})^T \right]^T . \tag{6}$$

Using the Euler-Bernoulli beam theory (i.e., the beam cross section remains planar and normal to the bean centerline), the virtual work of the elastic forces can be obtained as

$$\delta W^i = EA \int_{x^i} \delta\varepsilon^i \varepsilon^i dx^i + EI \int_{x^i} \delta\kappa^i \kappa^i dx^i = \delta\mathbf{e}^{iT} \mathbf{Q}_k^i , \tag{7}$$

where EA and EI are, respectively, the axial and flexural rigidity; ε^i and k^i are the axial strain and the curvature, respectively [26]. Notice here that this element does not account for pure torsion since rotation about the gradient vector is not considered. For this reason, this element is also called cable element [26,27]. Using the principle of virtual work in dynamics, the equations of motion of the ANCF beam element are written as

$$\mathbf{M}^i \ddot{\mathbf{e}}^i = \mathbf{Q}_k^i + \mathbf{Q}_e^i , \tag{8}$$

where $\mathbf{Q}_k^i$ is the vector of generalized element elastic forces, $\mathbf{Q}_e^i$ is the vector of generalized element external forces, and $\mathbf{M}^i$ is the constant element mass matrix defined by

$$\mathbf{M}^i = \int_{x^i} \rho^i A^i (\mathbf{S}^i)^T \mathbf{S}^i dx^i , \tag{9}$$

where ρ^i is the material density and A^i is the cross section area.

Thin Plate Element. There are two types of plate/shell elements in Chrono; the 4-node Kirchhoff thin plate element and the 4-node shear deformable shell element. In the thin plate element, the global position vector of a point in the middle plane of the plate element is defined by

$$\mathbf{r}^i = \mathbf{S}^i(x^i, y^i)\mathbf{e}^i , \tag{10}$$

where $\mathbf{S}^i(x^i, y^i)$ is the element shape function matrix obtained by the incomplete cubic polynomials [17,21]; x^i and y^i are the element coordinates in the middle plane. The nodal coordinates are defined as

$$\mathbf{e}^{ij} = \left[(\mathbf{r}^{ij})^T \quad \left(\frac{\partial \mathbf{r}^{ij}}{\partial x^i}\right)^T \quad \left(\frac{\partial \mathbf{r}^{ij}}{\partial y^i}\right)^T \right]^T . \tag{11}$$

Using the Kirchhoff-Love plate theory (i.e., the plate section remains planar and normal to the middle surface), the virtual work of the elastic forces can be obtained as [17,21]

$$\delta W^i = \int_{V^i} \delta\varepsilon^{iT} \mathbf{D}^i \varepsilon^i dV^i + \int_{A^i} \delta\kappa^{iT} \mathbf{D}_b^i \kappa^i dA^i = \delta\mathbf{e}^{iT} \mathbf{Q}_k^i , \tag{12}$$

where $\varepsilon^i = [\varepsilon_{xx}^i \quad \varepsilon_{yy}^i \quad \gamma_{xy}^i]^T$; $\kappa^i = [\kappa_{xx}^i \quad \kappa_{yy}^i \quad 2\kappa_{xy}^i]^T$; $\mathbf{D}^i$ and $\mathbf{D}_b^i$ are the elasticity matrices [17,21]. The in-plane strains and curvatures are defined as

$$\varepsilon_{xx}^i = \frac{1}{2} \left(\left(\frac{\partial \mathbf{r}^i}{\partial x^i}\right)^T \left(\frac{\partial \mathbf{r}^i}{\partial x^i}\right) - 1 \right)$$

$$\varepsilon_{yy}^i = \frac{1}{2} \left(\left(\frac{\partial \mathbf{r}^i}{\partial y^i}\right)^T \left(\frac{\partial \mathbf{r}^i}{\partial y^i}\right) - 1 \right) \tag{13}$$

$$\gamma_{xy}^i = \left(\frac{\partial \mathbf{r}^i}{\partial x^i}\right)^T \left(\frac{\partial \mathbf{r}^i}{\partial y^i}\right)$$

and

$$\kappa_{xx}^i = \mathbf{n}^{iT} \frac{\partial^2 \mathbf{r}^i}{\partial x^{i2}}, \quad \kappa_{yy}^i = \mathbf{n}^{iT} \frac{\partial^2 \mathbf{r}^i}{\partial y^{i2}}, \quad \kappa_{xy}^i = \mathbf{n}^{iT} \frac{\partial^2 \mathbf{r}^i}{\partial x^i \partial y^i}, \tag{14}$$

where the vector $\mathbf{n}^i$ is a unit normal to the middle plane defined by $(\frac{\partial \mathbf{r}^i}{\partial x^i} \times \frac{\partial \mathbf{r}^i}{\partial y^i})/|\frac{\partial \mathbf{r}^i}{\partial x^i} \times \frac{\partial \mathbf{r}^i}{\partial y^i}|$.

Shear Deformable Shell Element. In the shear deformable shell element that can be applied to thick shell structures with various material models, the global position vector of an arbitrary point in element i is defined by

$$\mathbf{r}^i = \mathbf{S}^i(x^i, y^i, z^i)\mathbf{e}^i , \tag{15}$$

where $\mathbf{S}^i$ is the element shape function matrix obtained by the bi-linear polynomials [16,92]. The preceding equation can be expressed as a sum of the displacement on the middle surface $\mathbf{r}_m^i$ and the displacement on the cross section as $\mathbf{r}^i = \mathbf{r}_m^i(x^i, y^i) + z^i \partial \mathbf{r}^i / \partial z^i(x^i, y^i)$. z^i is the element coordinate along the shell thickness. The nodal coordinates are defined as [16,92].

$$\mathbf{e}^{ij} = \left[(\mathbf{r}^{ij})^T \quad (\frac{\partial \mathbf{r}^{ij}}{\partial z^i})^T \right]^T . \tag{16}$$

The virtual work of the elastic forces can then be obtained as [92]

$$\delta W^i = \int_{V_0^i} \delta \varepsilon^{iT} \frac{\partial W^i(\hat{\varepsilon}^i)}{\partial \varepsilon^i} dV_0^i = \delta \mathbf{e}^{iT} \mathbf{Q}_k^i , \tag{17}$$

where dV_0^i is the infinitesimal volume at the initially curved reference configuration of element i, and W^i is an elastic energy density function. Due to the element lockings exhibited in this element, locking remedies need to be introduced to ensure the element convergence and accuracy. The lockings in the bi-linear shell element include the transverse shear locking; Poisson's thickness locking; curvature thickness locking; and in-plane shear locking. These lockings are systematically eliminated by applying the assumed natural strain method [9,11] and the enhanced assumed strain method [4,77]. By applying these locking remedies, the Green-Lagrange strain vector is defined as

$$\hat{\varepsilon}^i = (\mathbf{T}^i)^{-T} \tilde{\varepsilon}^i + \varepsilon^{i,EAS} , \tag{18}$$

where $\tilde{\varepsilon}^i$ is the covariant strain vector obtained from the covariant strain tensor given by:

$$\tilde{\mathbf{E}}^i = \frac{1}{2}((\bar{\mathbf{J}}^i)^T \bar{\mathbf{J}}^i - (\mathbf{J}^i)^T \mathbf{J}^i) . \tag{19}$$

In the preceding equation, $\bar{\mathbf{J}}^i = \partial \mathbf{r}^i / \partial \mathbf{x}^i$, $\mathbf{J}^i = \partial \mathbf{X}^i / \partial \mathbf{x}^i$ and $\mathbf{x}^i = [x^i \quad y^i \quad z^i]$, and $\mathbf{X}^i$ represents the global position vector of element i at an initially curved reference configuration. The transformation matrix $\mathbf{T}^i$ is as given in literature [92]. The assumed strain approach is introduced to the covariant transverse normal and transverse shear strains as follows:

$$\tilde{\varepsilon}^i = [\tilde{\varepsilon}_{xx} \quad \tilde{\varepsilon}_{yy} \quad \tilde{\gamma}_{xy} \quad \varepsilon_{zz}^{ANS} \quad \tilde{\gamma}_{xz}^{ANS} \quad \tilde{\gamma}_{yz}^{ANS}]^T . \tag{20}$$

The enhanced assumed strain method is applied to the in-plain strains and transverse normal strain as follows:

$$\varepsilon^{i,EAS} = [\varepsilon_{xx}^{EAS} \quad \varepsilon_{yy}^{EAS} \quad \gamma_{xy}^{EAS} \quad \varepsilon_{zz}^{EAS} \quad 0 \quad 0]^T . \tag{21}$$

It is important to note here that nonlinear constitutive models can be considered in a way same as solid elements. Using the principle of virtual work in dynamics, the equations of motion of the shear deformable shell element i can be expressed as

$$\mathbf{M}^i \ddot{\mathbf{e}}^i = \mathbf{Q}_k^i(\mathbf{e}^i, \boldsymbol{\alpha}^i) + \mathbf{Q}_e^i(\mathbf{e}^i, \dot{\mathbf{e}}^i, t) , \tag{22}$$

where vectors $\mathbf{Q}_k^i$ and $\mathbf{Q}_e^i$ are, respectively, vectors of the element elastic forces and external forces; and the matrix $\mathbf{M}^i$ is the constant element mass matrix defined by [92]

$$\mathbf{M}^i = \int_{V_0^i} \rho_0^i (\mathbf{S}^i)^T \mathbf{S}^i dV_0^i , \tag{23}$$

where ρ_0^{ik} is the material density at the reference configuration. The internal parameters α^i in Eq. 22, which are introduced to define the enhanced assumed strain field, are determined by solving the following equations [4, 77]

$$\int_{V_0^i} \left(\frac{\partial \varepsilon^{i,EAS}}{\partial \alpha^i} \right)^T \frac{\partial W^i(\hat{\varepsilon}^i)}{\partial \varepsilon^i} dV_0^i = \mathbf{0} . \qquad (24)$$

The equations above can be solved at element level for the unknown internal parameters using the procedure presented in the literature [92].

3.2 Nonlinear Finite Element Analysis via the Corotational Approach

The CR approach, see for instance [14, 15, 22], can be regarded as an augmentation of the classic linear finite element analysis, of whom it inherits the fundamental functions for computing the stiffness matrices and the internal forces. This fosters the reuse of finite element algorithms and theories whose behavior in the linear field are already well known and tested. Yet, the paradigm reuse comes at a cost: although displacements and rotations can be arbitrarily large, the CR framework requires that the strains must be small. In fact the CR concept is based on the idea that large deformations in beams, shells, etc., can be seen as local rigid body motions of finite elements, to whom small deformations can be superimposed - hence the possibility of using linear FEA formulations locally for the co-rotated elements.

Chrono uses the CR approach to model large deformations in meshes of 3D elements such as tetrahedrons and hexahedrons, as well as in beams discretized with classical elements such as Euler-Bernoulli beams. Figure 4 shows the concept of the corotational formulation as implemented in Chrono. Although the figure shows a beam, it applies equally well to tetrahedrons and other elements. We introduce an auxiliary floating coordinate system F per each element, and require that the coordinate system follows the deformed element. For a proper choice of F position update, the overall gross motion from the undeformed state into the deformed state C_D can be seen as the superposition of a large rigid body motion from the reference configuration C_0 to the so called *floating* or *shadow* configuration C_S, plus a local small-strain deformation from C_S to C_D.

The idea of the corotational approach is that one can compute the global tangent stiffness $\mathbf{K}_e$ and a global force $\mathbf{f}_e$ for each element e, given its local $\underline{K}$, its local $\underline{f}$ and the rigid body motion of the frame F in C_0 to F in C_S. When the element moves, the position and rotation of F is updated. In the case of beams, to avoid dependence on connectivity we place the origin of F in the midpoint of the AB segment, as $\mathbf{x}_F = \frac{1}{2}(\mathbf{x}_B - \mathbf{x}_A)$, and align its $\mathbf{X}$ axis with $\mathbf{x}_B - \mathbf{x}_A$. The remaining $\mathbf{Y}$ and $\mathbf{Z}$ axes of F are obtained via a Gram-Schmidt orthogonalization, enforcing $\mathbf{Y}$ to bisect the $\mathbf{Y}$ axes of A and B when projected on the plane orthogonal to $\mathbf{X}$. In case of tetrahedrons, the F frame is placed in the barycenter of the tetrahedron, and the alignment of the three axes is obtained using a polar decomposition that minimizes the displacement of the nodes with respect to the rotated F as described in [80].

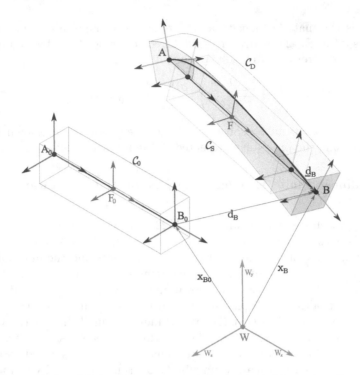

Fig. 4. The corotational finite element concept demonstrated in conjunction with a 3D beam element. For each finite element there is a floating frame F.

The matrix $\underline{K}$ and the vector $\underline{f}_{in}$ are evaluated using the classical theory for linear finite elements, since both are expressed in local coordinates. Then, both are transformed from the local to the global coordinates using the approach outlined in [22]. For the most part, this amounts to performing rotation transformations to the rows and columns corresponding to 3D nodes of $\underline{K}$ and $\underline{f}_{in}$, where the rotation matrix is the 3×3 matrix that contains the rotation of the floating coordinate system F. However, for completeness the mentioned approach also computes additional terms, especially the geometric stiffness matrix. Following [22,69] we also use projectors that filter the rigid body motion and that improve the consistency and the convergence of the method.

Figure 5 pertains one of the benchmarks performed to validate the methodology. This is the so called Princeton beam experiment for which ample experimental results are available in the literature [18,19]. In this numerical experiment thin beams were each modeled with ten Euler-Bernoulli corotational beam elements. A force was applied to the tip, with increasing magnitude and inclination with respect to the vertical. Because of the diagonal nature of the force, beams are bent and slightly twisted. If a simple linear analysis were used, the twisting effect would not take place. Results, presented in Fig. 6, show that there is good agreement with experimental results and with other third party software; i.e., Dymore and MBDyn [10,46].

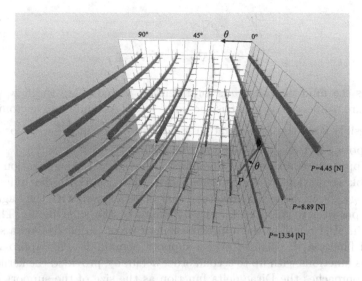

Fig. 5. Non-linear static analysis of the Princeton beam experiment, at different load magnitudes and θ angles.

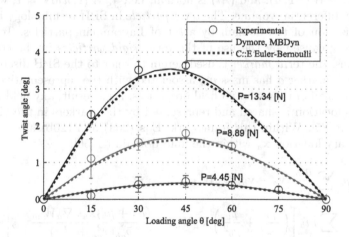

Fig. 6. Results from the non-linear analysis of the Princeton beam experiment (Color figure online).

4 Fluid–Solid Interaction Support in **Chrono**

In a Lagrangian framework, the continuity and momentum equations associated with the fluid dynamics assume the form of Ordinary Differential Equations (ODE) [45]

$$\frac{D\rho}{dt} = -\rho \nabla \cdot \mathbf{v} \tag{25a}$$

$$\frac{D\mathbf{v}}{dt} = -\frac{1}{\rho}\nabla p + \frac{\mu}{\rho}\nabla^2 \mathbf{v} + \mathbf{f} , \tag{25b}$$

where μ is the fluid viscosity, ρ is fluid density, $\mathbf{v}$ and p are the flow velocity and pressure, respectively, and $\mathbf{f}$ is the body force. Assuming a Newtonian and incompressible flow, the first equation translates either in $\dfrac{D\rho}{dt} = 0$, or equivalently, imposing a divergence-free flow condition, in $\nabla \cdot \mathbf{v} = 0$.

The approach embraced in Chrono for the spatial discretization of the Navier-Stokes equations draws on the Smoothed Particle Hydrodynamics (SPH) methodology [29, 42], a meshless method that dovetails well with the Lagrangian modeling perspective adopted for the dynamics of the solid phase. The term *smoothed* in SPH refers to the approximation of point properties via a smoothing kernel function W, defined over a support domain S. This approximation reproduces functions with up to second order accuracy, provided the kernel function: (*i*) approaches the Dirac delta function as the size of the support domain tends to zero, that is $\lim_{h\to 0} W(\mathbf{r}, h) = \delta(\mathbf{r})$, where $\mathbf{r}$ is the spatial distance and h is a characteristic length that defines the kernel smoothness; (*ii*) is symmetric, i.e., $W(\mathbf{r}, h) = W(-\mathbf{r}, h)$; and (*iii*) is normal, i.e., $\int_S W(\mathbf{r}, h)d\mathbb{V} = 1$, where $d\mathbb{V}$ denotes the differential volume. The term *particle* in SPH terminology indicates the discretization of the domain by a set of Lagrangian particles. To remove the ambiguity caused by the use of the term *rigid particles* in the context of FSI problems, the term *marker* is used herein to refer to the SPH discretization process. Each marker a has mass m_a associated with the representative volume $d\mathbb{V}$ and carries all of the essential field properties. As a result, any field property at a certain location is shared and represented by the markers in the vicinity of that location [85]. Within this framework, Eqs. (25a) and (25b) are discretized at an arbitrary location $\mathbf{x}_a$ within the fluid domain as [56]:

$$\frac{d\rho_a}{dt} = \rho_a \sum_b \frac{m_b}{\rho_b} (\mathbf{v}_a - \mathbf{v}_b) \cdot \nabla_a W_{ab} \tag{26a}$$

$$\frac{d\mathbf{v}_a}{dt} = -\sum_b m_b \left[\left(\frac{p_a}{\rho_a{}^2} + \frac{p_b}{\rho_b{}^2} \right) \nabla_a W_{ab} - \frac{(\mu_a + \mu_b)\mathbf{x}_{ab} \cdot \nabla_a W_{ab}}{\bar{\rho}_{ab}^2 (x_{ab}^2 + \varepsilon h^2)} \mathbf{v}_{ab} \right] + \mathbf{f}_a , \tag{26b}$$

which are followed by

$$\frac{d\mathbf{x}_a}{dt} = \mathbf{v}_a , \tag{27}$$

to update the location of discretization markers. In the above equations, quantities with subscripts a and b are associated with markers a and b, respectively; $\mathbf{x}_{ab} = \mathbf{x}_a - \mathbf{x}_b$, $\mathbf{v}_{ab} = \mathbf{v}_a - \mathbf{v}_b$, $W_{ab} = W(\mathbf{x}_{ab}, h)$, $\bar{\rho}_{ab}$ is the average density of markers a and b, ∇_a is the gradient with respect to $\mathbf{x}_a$, i.e., $\partial/\partial \mathbf{x}_a$, and ε is a regularization coefficient to prevent infinite reaction between markers sharing the same location.

The current SPH implementation in Chrono relies on a weakly compressible model, namely Eq. (26a) followed by a equation of state to update pressure p [55]:

$$p = \frac{c_s^2 \rho_0}{\gamma} \left\{ \left(\frac{\rho}{\rho_0} \right)^\gamma - 1 \right\}. \tag{28}$$

In this equation, ρ_0 is the reference density of the fluid, γ tunes the stiffness of the pressure-density relationship, and c_s is the speed of sound. The value c_s is adjusted depending on the maximum speed of the flow, $V_{\max}$, to keep the flow compressibility below some arbitrary value. Typically, $\gamma = 7$ and $c_s = 10V_{\max}$, which allows 1 % flow compressibility [55].

Two additional refinements are adopted to improve the convergence properties of the weakly compressible SPH implementation. In the first adjustment, an extended SPH approach (XSPH) [54] is employed to modify the individual markers velocity based on collective; i.e., Eulerian, velocities. This regularization prevents excessive marker overlaps. The second modification is a re-initialization technique [13] which ensures consistency between marker densities updated through Eq. (26a) and those obtained directly from $\rho_a = \sum_b m_b W_{ab}$. The implementation details and algorithms can be found in [65, 66].

Several methods are proposed to enforce a fixed or moving solid boundary [3, 8, 36, 56, 65]. In our previous work, we used an approach based on so-called Boundary Condition Enforcing (BCE) markers, which are distributed on the rigid [65] or flexible [67] bodies to capture the FSI interactions. At the fluid-solid interface, each BCE marker captures an interaction force due to its inclusion in the proximity of the nearby fluid markers through Eq. (26). Two approaches were implemented to update the velocity and pressure of a BCE marker. In the first approach, the marker velocity is replaced by the local velocity of the

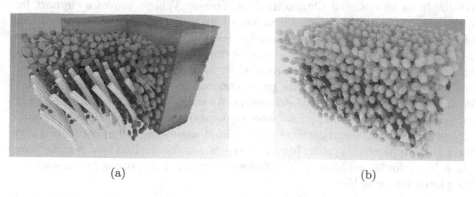

(a)   (b)

Fig. 7. GPU simulation of the flow of rigid bodies within an array of flexible beams. For a clear visualization, only parts of the domain are shown in each picture: (a) rigid bodies, flexible beams, and fluid flow; (b) rigid bodies and flexible beams only (fluid not rendered). The velocity field is color-coded: from zero (blue) to maximum (red), with $V_{max}^{fluid} = 0.045$ m/s, $V_{max}^{rigid} = 0.041$ m/s, and $V_{max}^{beam} = 0.005$ m/s (Color figure online).

moving boundary, while the pressure relies on a projection from the fluid domain [41]. In the second approach, the velocity of each BCE marker is calculated so that when combined with the fluid contribution, it results in the assigned wall velocity; additionally, BCE pressure is obtained from a force balance at the boundary. We showed that the second approach performs better in imposing the no-slip condition, particularly when the external body force is significant [68].

To achieve the computational efficiency required for complex multi-physics simulations, we adopted a parallel programming approach relying on GPU computing [67]. The resulting computational framework has been leveraged to investigate FSI problems in rigid body/particle suspension [65], three-way interaction of fluid, rigid, and flexible components [67], and microfluidic sorting of microtissues [64]. Figure 7 shows a snapshot of a simulation that involves immersed rigid and flexible components, where an ANCF method (see Sect. 3.1) is used to simulate the flexible beams.

5   Chrono Toolkits

A Chrono toolkit is a set of pre- and post-processing utilities that encapsulate domain expertise and provide a low entry point to Chrono by capturing expert knowledge in ready-to-use generic modeling templates in a focused research/application area. For instance, Chrono::Vehicle provides a collection of subsystem templates that can be quickly used to assemble a typical vehicle.

Chrono::Vehicle and Chrono::Granular are currently available and discussed herein. Chrono::Robotics and Chrono::Terramechanics are being developed.

5.1   Chrono::Vehicle

Available as an optional Chrono module, Chrono::Vehicle provides support for modeling, simulation, and visualization of ground vehicle systems. Modeling of vehicle systems is done in a modular fashion, with a vehicle defined as an assembly of instances of various subsystems (suspension, steering, driveline, etc.). Flexibility in modeling is provided by adopting a template-based design. In Chrono::Vehicle templates are parameterized models that define a particular implementation of a vehicle subsystem. As such, a template defines the basic modeling elements (bodies, joints, force elements), imposes the subsystem topology, prescribes the design parameters, and implements the common functionality for a given type of subsystem (e.g., suspension) particularized to a specific template (e.g., double wishbone). The following vehicle subsystems and associated templates are available:

Suspension: double wishbone, reduced double wishbone (with the A-arms modeled as distance constraints), multi-link, solid-axle, walking-beam;
Steering: Pitman arm, rack-and-pinion;
Anti-roll bar: two-body, rotational spring-damper-based anti-roll bar;

Driveline: 2WD shaft-based, 4WD shaft-based; these templates are based on specialized Chrono modeling elements, named `ChShaft`, with a single rotational degree of freedom and various shaft coupling elements (gears, differentials, etc.);

Wheel: in Chrono::Vehicle, a wheel only carries additional mass and inertia appended to the suspension's spindle body and, optionally, visualization information;

Brake: simple brake (constant torque modulated by the driver braking input).

Figure 8 illustrates the representation of a wheeled vehicle as an assembly of instances of various subsystem templates (here using double wishbone suspensions both in the front and rear and a Pitman-arm steering mechanism).

For additional flexibility and to facilitate inclusion in larger simulation frameworks, Chrono::Vehicle allows formally separating various systems (the vehicle itself, powertrain, tires, terrain, driver) and provides the inter-system communication API for a co-simulation framework based on force-displacement couplings. For consistency, these systems are themselves templatized:

Vehicle: a collection of references used to instantiate templates for its constitutive subsystems;

Powertrain: shaft-based template using an engine model based on speed-torque curves, torque converter based on capacity factor and torque ratio curves, and transmission parameterized by an arbitrary number of forward gear ratios and a single reverse gear ratio;

Tire: rigid tire (based on the Chrono rigid contact model), Pacejka, and a LuGre friction tire model;

Driver: interactive driver model (with user inputs from keyboard for real-time simulation), file-based driver model (interpolated driver inputs as functions of time).

Tire models supported in Chrono::Vehicle can be broken down into two categories: (*i*) a set of well developed traditional models, and (*ii*) a FEA-based tire model actively under development. In the first category, three tire models are currently supported: a rigid tire model, a tire model based on the LuGre friction model, and single contact point Pacejka tire model. The rigid tire model leverages the collision detection and frictional contact support in Chrono by using tires with user-specified geometry and contact material properties which are then treated like any other collision geometry within Chrono. The LuGre tire model described in [53] is based on a lumped brush-based LuGre friction model. The tire is broken down into a user specified number of identical, connected two dimensional disks. For each disk, a normal force is calculated based on the disk–ground penetration distance and velocity; the lateral and longitudinal forces are calculated by integrating the differential equations for the friction force in each direction with no coupling between the lateral and longitudinal directions. The forces are then summed across all the disks and the resulting force is then transformed and applied to the wheel center. The third traditional tire model is a Pacejka-based formulation described in [44]. This model is a modification

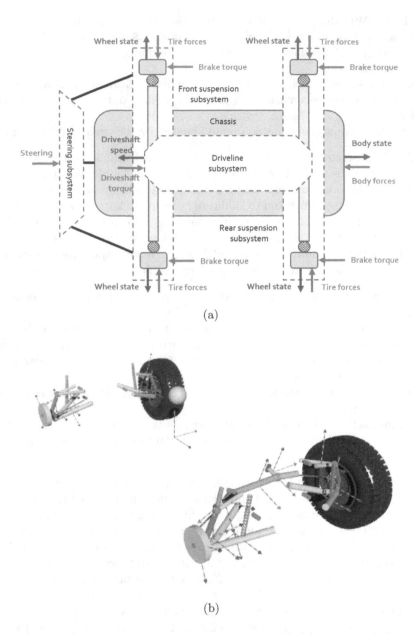

(a)

(b)

Fig. 8. A two-axle, independent suspension vehicle model: (a) diagram of the component subsystems in the vehicle assembly; (b) a possible realization in Chrono::Vehicle (in this particular case, using double wishbone suspensions, a Pitman-arm steering mechanism, and a 4WD driveline subsystem – the latter has no graphical representation in Chrono::Vehicle). The image on the left also illustrates the data exchange between the vehicle system and associated systems (powertrain, tires, etc.)

of the 2013 ADAMS/Tire PAC2002 non-linear transient tire model without belt dynamics, valid for tire responses up to approximately 15 Hz. In the second category, a high-fidelity deformable tire model based on ANCF (see Sect. 3.1) is implemented in Chrono. The fiber reinforced rubber material of tires is modeled by the laminated composite shell element, and the distributed parameter LuGre tire friction model is utilized to allow for an accurate prediction of the shear contact stress distribution over the contact patch under various vehicle maneuvering scenarios [84].

For easy incorporation in and interoperability with third-party applications, Chrono::Vehicle is provided as a middleware library. System and subsystem templates are implemented through polymorphic classes and thus the library is extensible through C++ inheritance, allowing definition of new subsystems or new templates for existing subsystems. Systems and subsystems are defined with a three-layered C++ class hierarchy: (i) base abstract class for type (e.g., ChSuspension); (ii) derived, still abstract, class for specific template (e.g., ChDou-bleWishbone); and (iii) concrete class that particularizes the template for a given vehicle (e.g., HMMWV_DoubleWishboneFront). We provide two different types for the concrete classes that specify a template. The first one, providing maximum flexibility, is through *user-defined* classes which implement all virtual functions imposed by the corresponding template base class. While not part of the Chrono::Vehicle library itself, several examples of this approach are provided with the package. A more convenient approach, allowing for fast experimentation and parametric studies, is offered through a set of concrete template classes (part of the Chrono::Vehicle library) that specify the subsystem for a particular vehicle through specification data files in the JSON format [1].

Visualization of vehicle systems is provided both for run-time, interactive simulation (through the Chrono built-in Irrlicht visualization support) and for high quality post-processing rendering (using for example the POV-Ray ray-tracing package, see Fig. 9).

Currently only wheeled vehicles are supported, but work is underway for extending Chrono::Vehicle to include templates appropriate for modeling tracked vehicles.

5.2   Chrono::Granular

To facilitate the accurate modeling of particulate or granular materials used in Chrono simulations, a module called Chrono::Granular is available. This module provides pre-processing and post-processing tools specific to granular materials, as well as C++ templates for a variety of material tests, including the standard geomechanics tests (e.g., direct shear, triaxial) to determine elastic and plastic properties of the bulk granular material, as well as standard industrial processes (e.g., hopper flow, conveyor transport). Chrono::Granular can be used on its own or in conjunction with other modules, such as Chrono::Vehicle, e.g., in scenarios where vehicle-ground interactions are critical and where the ground is a granular material, such as sand, soil, or gravel.

Fig. 9. A Chrono::Vehicle simulation of a HMMWV vehicle operating on granular terrain composed of over 150000 bodies. Each granular particle is an ellipsoid that can be circumscribed in a sphere with a radius of 2 cm. The solver uses the complementarity form of contact with a time step of 0.001 s and 40 APGD iterations per step. The simulation took approximately 3.2 computation hours per second on a 3 GHz Intel Xeon E5-2690v2 processor using 20 cores.

The pre-processing tools provided by Chrono::Granular include the specification of granular materials with either user-definable or preset particle size and shape distributions, e.g., for ASTM standard graded sand; and various bulk geometries, e.g., cubical for the direct shear box test or cylindrical for the standard triaxial test. Once specified, granular material specimens can be subjected to the standard tests provided by the Chrono::Granular templates, to determine bulk elastic and plastic properties, such as Young's modulus, Poisson's ratio, and the friction and dilation angles of the bulk granular material. These bulk material properties for the simulated granular material are extracted from the simulations by the templates, and they can be compared to the bulk physical properties of the granular material that the user wishes to model for validation before subsequent Chrono simulations are run.

In addition to the determination of macro-scale or bulk granular material properties, Chrono::Granular also provides post-processing tools for the determination and visualization of micro-scale or local granular material properties, including particle trajectories, inter-particle force chains, and the local stress distribution [60], defined as $\sigma_{ij} = \left(\sum_c f_i^c l_j^c\right)/V_\sigma$, where σ_{ij} is the average local stress tensor over a representative volume including at least two particles, f_i^c is the contact force vector, and l_j^c is the branch vector connecting the centers of contacting pairs, as shown in Fig. 10. The sum is over the contacts between the particles within the representative volume, and V_σ is the volume of the region containing those particles.

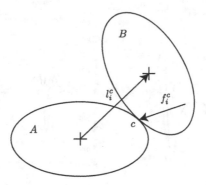

Fig. 10. An inter-particle contact c between two particles within a granular material, with the branch vector l_i^c and the contact force vector f_i^c shown.

6 Chrono Validation Studies

Validation Against ADAMS. To validate basic modeling elements, a series of systems were created within Chrono. Multiple models were generated to test different aspects of each modeling element examined: revolute, spherical, universal, prismatic, and cylindrical joints; distance and revolute-spherical joints; translational and rotational spring-dampers; and translational actuators. Since the majority of the basic test systems did not have closed form analytical solutions, equivalent models were constructed and solved within MSC.ADAMS. The simulation results between Chrono and ADAMS were compared for each model to ensure close agreement between the two programs. For all of the test models examined, the results between the simulation programs were in close alignment as described in [88]. As an example, consider one of the unit tests for the universal joint. In this test, a pendulum body, initially at rest and aligned along the $x = y$ line, is connected to the ground through a universal joint; the initial directions of the universal joint's cross are $[1/2, -1/2, \sqrt{2}/2]$ and $[-1/2, 1/2, \sqrt{2}/2]$. The comparison plots in Fig. 11 show that the maximum difference in angular acceleration between Chrono and ADAMS is on the order of 0.5 % of the peak angular acceleration seen in the system. Although not shown, the results between Chrono and ADAMS would converge further if the solver tolerances were tightened beyond the settings used in this study.

Validation Against Experimental Data. The DEM-P and DEM-C contact approaches were validated against several fundamental terramechanics experiments. The first validation test, shown in Fig. 12(a), used an aluminum rig designed and fabricated to measure the gravity-induced mass flow rate through a gap of a specified amount of granular material. The flow of the 500 micron diameter glass spheres was simulated for each gap size used in the lab experiments. Figure 12(b) shows the simulation results plotted next to experimental measurements (weight as a function of time). Good statistical correlation was observed between simulation and experimental results for both DEM-P and DEM-C.

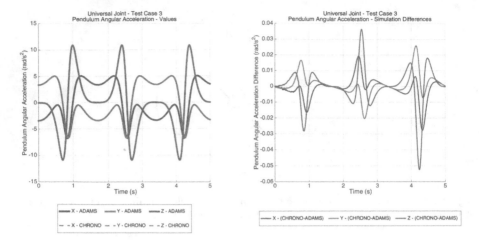

Fig. 11. Comparison of the pendulum angular acceleration in the global frame between CHRONO and ADAMS for the universal joint test system 3 (Color figure online)

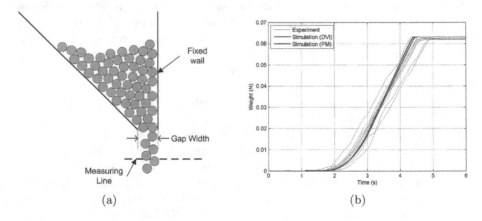

(a)   (b)

Fig. 12. Schematic of mass flow rate validation experiment in the open configuration (left). Weight of collected granular material vs. time for 2.0 mm gap size (right) (Color figure online).

A second validation test, shown in Fig. 13, was run for an impact problem in a series of simulations reported in [34]. A relation of the form

$$d = \frac{0.14}{\mu} \left(\frac{\rho_b}{\rho_g}\right)^{1/2} D_b^{2/3} H^{1/3} \tag{29}$$

has been empirically established [90], where d is the depth of penetration, h is the height from which a ball of density ρ_b and diameter D_b is dropped, $H = d + h$ is the total drop distance, and ρ_g is the density of the granular material. Finally, μ was the friction coefficient in the granular material obtained from an angle

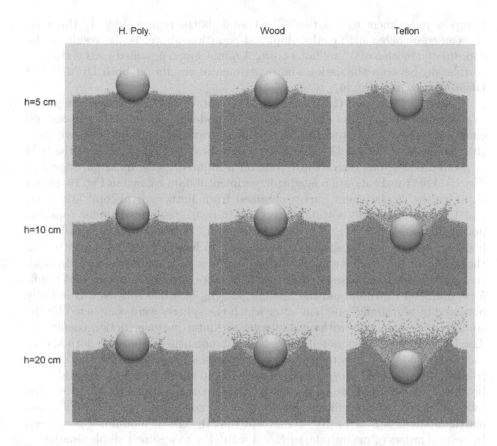

Fig. 13. Snapshot of the instant of deepest penetration from each impact simulation.

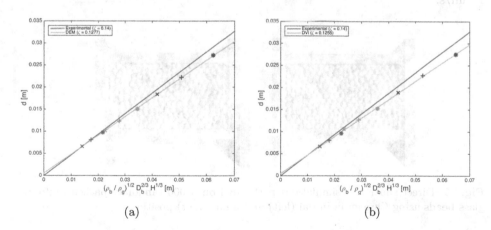

Fig. 14. Penetration depth vs. scaled total drop distance for the DEM-P (left) and DEM-C (right) contact methods (Color figure online).

of repose experiment as $\mu = \tan(\theta_r)$, where θ_r is the repose angle. In this test, the emergent behavior, i.e. the empirical equation above, is the result of the coordinated motion of 0.5 million bodies. Again, Chrono provided good statistical correlation between simulation and experimental results for both DEM-P and DEM-C, shown in Fig. 14.

Lastly, to verify that the Chrono DEM-P contact model with true tangential displacement history does indeed accurately model the micro-scale physics and emergent macro-scale properties of a simple granular material, we have used the templates of Chrono::Granular to simulate physical tests typical of the field of geomechanics; i.e., direct shear tests on a mono-disperse material, shown in Fig. 15. This third validation against experimental data, shown in Fig. 16, shows shear versus displacement curves obtained from both experimental [32] (top) and Chrono-simulated (bottom) direct shear tests, performed under constant normal stresses of 3.1, 6.4, 12.5, and 24.2 kPa, on 5,000 uniform glass beads. The inside dimensions of the shear box were 12 cm in length by 12 cm in width, and the height of the granular material specimen was approximately 6 cm. In both the experimental and simulated direct shear tests, the glass spheres had a uniform diameter of 6 mm, and the random packing of 5,000 spheres was initially obtained by a "rainfall" method, after which the spheres were compacted by the confining normal stress without adjusting the inter-particle friction coefficient. The DEM-P simulations were performed in Chrono using a Hertzian normal contact force model and true tangential contact displacement history with Coulomb friction. The material properties of the spheres in the simulations were taken to be those corresponding to glass [32], for which the density is 2,550 kg/m³, the inter-particle friction coefficient is $\mu = 0.18$, Poisson's ratio is $\nu = 0.22$, and the elastic modulus is $E = 4 \times 10^{10}$ Pa, except that the elastic modulus was reduced by several orders of magnitude, to $E = 4 \times 10^7$ Pa, to ensure a stable simulation with a reasonable time integration step-size of $h = 10^{-5}$ s. The shear speed was 1 mm/s.

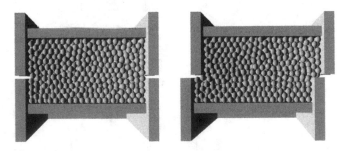

Fig. 15. Direct shear test simulations performed on 5,000 randomly packed uniform glass beads using Chrono, in initial (left) and final (right) positions.

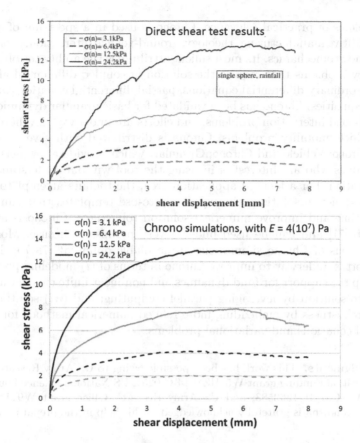

Fig. 16. Direct shear test results for 5,000 randomly packed uniform glass beads, obtained by experiment [32] (top) and DEM-P simulations using Chrono (bottom), under constant normal stresses of 3.1, 6.4, 12.5, and 24.2 kPa.

From Fig. 16, one can calculate the friction angle ϕ for the bulk granular material, which is the inverse tangent of the ratio of shear stress to normal stress at the initiation of yield (the peak friction angle ϕ_p) and post yield (the residual friction angle ϕ_r) during the direct shear test [7]. Specifically, for constant normal stresses of 3.1, 6.4, 12.5, and 24.2 kPa, the peak friction angles ϕ_p are approximately 33, 32, 29, and 28 deg., respectively, for both the experimental and Chrono-simulated direct shear tests.

7   Conclusions and Future Work

Chrono is an ongoing open source software development project aimed at establishing a dynamics engine that is experimentally validated and which draws on parallel computing and a broad spectrum of modeling techniques to solve large problems of practical relevance. Chrono is used in a spectrum of applications in additive manufacturing, rheology, ground-vehicle interaction, soft-matter physics, and geomechanics. Its most salient attribute is the ability to solve multi-disciplinary problems that require the solution of coupled differential algebraic equations, ordinary differential equations, partial differential equations and variational inequalities. Chrono has been validated for basic granular dynamics problems, fluid-solid interaction problems, and efforts are underway to validate it for ground-vehicle mobility problems. Chrono is distributed with several toolkits, such as Chrono::Vehicle and Chrono::Granular, which provide a low entry point for individuals who are interested in using the tool without understanding its implementation. For a focused application area, the toolkits attempt to encapsulate best-practice solutions through ready-to-use templates that can reduce modeling time and improve numerical solution robustness. Chrono is available on GitHub [71] and can be used under a very permissive license. More than 150 animations of Chrono simulations are available online at [78,79]. Looking ahead, effort is underway to improve Chrono in terms of (i) modeling prowess by building up the support for fluid dynamics and nonlinear finite element analysis; (ii) time to solution by leveraging parallel computing; and (iii) solution accuracy and robustness by embedding more refined numerical methods for solving differential equations and variational problems.

Acknowledgments. This work has been possible owing to US Army Research Office Rapid Innovation Funding grant W56HZV-14-C-0254, US National Science Foundation grant GOALI-CMMI 1362583, and US Army Research Office grant W911NF-12-1-0395. Milad Rakhsha is gratefully acknowledged for his help in the preparation of this manuscript.

References

1. The JSON data interchange format. Technical report ECMA-404, ECMA International (2013)
2. Acary, V., Brogliato, B.: Numerical Methods for Nonsmooth Dynamical Systems: Applications in Mechanics and Electronics, vol. 35. Springer Science & Business Media, Heidelberg (2008)
3. Adami, S., Hu, X., Adams, N.: A generalized wall boundary condition for smoothed particle hydrodynamics. J. Comput. Phys. **231**, 7057–7075 (2012)
4. Andelfinger, U., Ramm, E.: EAS-elements for two-dimensional, three-dimensional, plate and shell structures and their equivalence to HR-elements. Int. J. Numer. Meth. Eng. **36**, 1311–1337 (1993)
5. Anitescu, M., Cremer, J.F., Potra, F.A.: Formulating 3D contact dynamics problems. Mech. Struct. Mach. **24**(4), 405–437 (1996)
6. Anitescu, M., Tasora, A.: An iterative approach for cone complementarity problems for nonsmooth dynamics. Comput. Optim. Appl. **47**, 207–235 (2010)
7. Bardet, J.-P.: Experimental Soil Mechanics. Prentice Hall, Englewood Cliffs (1997)
8. Basa, M., Quinlan, N., Lastiwka, M.: Robustness and accuracy of SPH formulations for viscous flow. Int. J. Numer. Meth. Fluids **60**, 1127–1148 (2009)

9. Bathe, K.-J., Dvorkin, E.N.: A four-node plate bending element based on mindlin/reissner plate theory and a mixed interpolation. Int. J. Numer. Meth. Eng. **21**, 367–383 (1985)
10. Bauchau, O.A.: DYMORE user's manual. Georgia Institute of Technology, Atlanta (2007)
11. Betsch, P., Stein, E.: An assumed strain approach avoiding artificial thickness straining for a non-linear 4-node shell element. Commun. Numer. Methods Eng. **11**, 899–909 (1995)
12. Buildbot: Buildbot - an open-source framework for automating software build, test, and release. http://buildbot.net/. Accessed 31 May 2015
13. Colagrossi, A., Landrini, M.: Numerical simulation of interfacial flows by smoothed particle hydrodynamics. J. Comput. Phys. **191**, 448–475 (2003)
14. Crisfield, M.: A consistent co-rotational formulation for non-linear, three-dimensional, beam-elements. Comput. Methods Appl. Mech. Eng. **81**, 131–150 (1990)
15. Crisfield, M.A., Galvanetto, U., Jelenic, G.: Dynamics of 3-D co-rotational beams. Comput. Mech. **20**, 507–519 (1997)
16. Dmitrochenko, O., Matikainen, M., Mikkola, A.: The simplest 3-and4-noded fully-parameterized ANCF plate elements. In: ASME 2012 International Design Engineering Technical Conferences and Computers and Information in Engineering Conference, American Society of Mechanical Engineers, pp. 317–322 (2012)
17. Dmitrochenko, O.N., Pogorelov, D.Y.: Generalization of plate finite elements for absolute nodal coordinate formulation. Multibody Sys.Dyn. **10**, 17–43 (2003)
18. Dowell, E.H., Traybar, J.J.: An experimental study of the nonlinear stiffness of a rotor blade undergoing flap, lag, and twist deformations, Aerospace and Mechanical Science Report 1194, Princeton University, January 1975
19. Dowell, E.H., Traybar, J.J.: An experimental study of the nonlinear stiffness of a rotor blade undergoing flap, lag, and twist deformations, Aerospace and Mechanical Science Report 1257, Princeton University, December 1975
20. Doxygen: Doxygen - A Documentation Generator From Annotated C++ Code. http://www.doxygen.org. Accessed 31 May 2015
21. Dufva, K., Shabana, A.: Analysis of thin plate structures using the absolute nodal coordinate formulation. Proc. Inst. Mech. Eng. Part K: J. Multi-body Dyn. **219**, 345–355 (2005)
22. Felippa, C., Haugen, B.: A unified formulation of small-strain corotational finite elements: I. theory. Comput. Methods Appl. Mech. Eng. **194**, 2285–2335 (2005). Computational Methods for Shells
23. Filippov, A.F., Arscott, F.M.: Differential Equations with Discontinuous Right-hand Sides: Control Systems, vol. 18. Springer, Heidelberg (1988)
24. Fleischmann, J.: DEM-PM contact model with multi-step tangential contact displacement history. Technical report TR-2015-06, Simulation-Based Engineering Laboratory, University of Wisconsin-Madison (2015)
25. Fleischmann, J.A., Serban, R., Negrut, D., Jayakumar, P.: On the importance of displacement history in soft-body contact models. ASME J. Comput. Nonlinear Dyn. (2015). doi:10.1115/1.4031197
26. Gerstmayr, J., Shabana, A.: Analysis of thin beams and cables using the absolute nodal co-ordinate formulation. Nonlinear Dyn. **45**, 109–130 (2006)
27. Gerstmayr, J., Sugiyama, H., Mikkola, A.: Review on the absolute nodal coordinate formulation for large deformation analysis of multibody systems. ASME J. Comput. Nonlinear Dyn. **8**, 031016-1–031016-12 (2013)

28. Gilardi, G., Sharf, I.: Literature survey of contact dynamics modelling. Mech. Mach. Theor. **37**, 1213–1239 (2002)
29. Gingold, R.A., Monaghan, J.J.: Smoothed particle hydrodynamics-theory and application to non-spherical stars. Mon. Not. R. Astron. Soc. **181**, 375–389 (1977)
30. Glocker, C., Pfeiffer, F.: An LCP-approach for multibody systems with planar friction. In: Proceedings of the CMIS 92 Contact Mechanics Int. Symposium, Lausanne, Switzerland, pp. 13–20 (2006)
31. Hairer, E., Wanner, G.: Solving Ordinary Differential Equations II: Stiff and Differential-Algebraic Problems. Springer, Heidelberg (1996)
32. Hartl, J., Ooi, J.: Experiments and simulations of direct sheartests: porosity, contact friction and bulk friction. Granular Matter **10**, 263–271 (2008)
33. Haug, E.J.: Computer-Aided Kinematics and Dynamics of Mechanical Systems Volume-I. Prentice-Hall, Englewood Cliffs (1989)
34. Heyn, T.: On the modeling, simulation, and visualization of many-body dynamics problems with friction and contact. Ph.D. thesis, Department of Mechanical Engineering, University of Wisconsin–Madison (2013). http://sbel.wisc.edu/documents/TobyHeynThesis_PhDfinal.pdf
35. Hindmarsh, A., Brown, P., Grant, K., Lee, S., Serban, R., Shumaker, D., Woodward, C.: SUNDIALS: suite of nonlinear and differential/algebraic equation solvers. ACM Trans. Math. Softw. (TOMS) **31**, 363–396 (2005)
36. Hu, W., Tian, Q., Hu, H.: Dynamic simulation of liquid-filled flexible multibody systems via absolute nodal coordinate formulation and SPH method. Nonlinear Dyn. **75**, 653–671 (2013)
37. Kaufman, D.M., Pai, D.K.: Geometric numerical integration of inequality constrained. SIAM J. Sci. Comput. Nonsmooth Hamiltonian Syst. **34**, A2670–A2703 (2012)
38. Kitware: CMake – A cross-platform, open-source build system. http://www.cmake.org. Accessed 31 May 2015
39. Kruggel-Emden, H., Simsek, E., Rickelt, S., Wirtz, S., Scherer, V.: Review and extension of normal force models for the discrete element method. Powder Technol. **171**, 157–173 (2007)
40. Kruggel-Emden, H., Wirtz, S., Scherer, V.: A study of tangential force laws applicable to the discrete element method (DEM) for materials with viscoelastic or plastic behavior. Chem. Eng. Sci. **63**, 1523–1541 (2008)
41. Lee, E., Moulinec, C., Xu, R., Violeau, D., Laurence, D., Stansby, P.: Comparisons of weakly compressible and truly incompressible algorithms for the SPH mesh free particle method. J. Comput. Phys. **227**, 8417–8436 (2008)
42. Lucy, L.B.: A numerical approach to the testing of the fission hypothesis. Astron. J. **82**, 1013–1024 (1977)
43. Machado, M., Moreira, P., Flores, P., Lankarani, H.M.: Compliant contact force models in multibody dynamics: evolution of the Hertz contact theory. Mech. Mach. Theor. **53**, 99–121 (2012)
44. Madsen, J.: Validation of a single contact point tire model based on the transient pacejka model in the open-source dynamics software chrono. Technical report, University of Wisconsin - Madison Simulation Based Engineering Lab (2014)
45. Malvern, L.E.: Introduction to the Mechanics of a Continuous Medium. Prentice Hall, Englewood Cliffs (1969)
46. Masarati, P., Morandini, M., Quaranta, G., Mantegazza, P.: Computational aspects and recent improvements in the open-source multibody analysis software MBDyn. In: Multibody Dynamics, pp. 21–24 (2005)

47. Mazhar, H., Bollmann, J., Forti, E., Praeger, A., Osswald, T., Negrut, D.: Studying the effect of powder geometry on the selective laser sintering process. In: Society of Plastics Engineers (SPE) ANTEC (2014)
48. Mazhar, H., Heyn, T., Pazouki, A., Melanz, D., Seidl, A., Bartholomew, A., Tasora, A., Negrut, D.: Chrono: a parallel multi-physics library for rigid-body, flexible-body, and fluid dynamics. Mech. Sci. **4**, 49–64 (2013)
49. Mazhar, H., Heyn, T., Tasora, A., Negrut, D.: Using Nesterov's method to accelerate multibody dynamics with friction and contact. ACM Trans. Graph. **34**, 32 (2015)
50. Mazhar, H., Osswald, T., Negrut, D.: On the use of computational multibody dynamics for increasing throughput in 3D printing. Addit. Manuf. (2016, accepted)
51. Melanz, D.: On the validation and applications of a parallel flexible multi-body dynamics implementation. M.S. thesis, University of Wisconsin-Madison (2012)
52. Melanz, D., Tupy, M., Smith, B., Turner, K., Negrut, D.: On the validation of a differential variational inequality approach for the dynamics of granular material-DETC2010-28804. In: Fukuda, S., Michopoulos, J.G. (eds.) Proceedings to the 30th Computers and Information in Engineering Conference, ASME International Design Engineering Technical Conferences (IDETC) and Computers and Information in Engineering Conference (CIE) (2010)
53. Mikkola, A.: Lugre tire model for HMMWV. Technical report TR-2014-15, Simulation-Based Engineering Laboratory, University of Wisconsin-Madison (2014)
54. Monaghan, J.J.: On the problem of penetration in particle methods. J. Comput. Phys. **82**, 1–15 (1989)
55. Monaghan, J.J.: Smoothed particle hydrodynamics. Rep. Prog. Phys. **68**, 1703–1759 (2005)
56. Morris, J.P., Fox, P.J., Zhu, Y.: Modeling low Reynolds number incompressible flows using SPH. J. Comput. Phys. **136**, 214–226 (1997)
57. MPICH2: High Performance Portable MPI (2013). http://www.mpich.org/
58. Negrut, D., Heyn, T., Seidl, A., Melanz, D., Gorsich, D., Lamb, D.: Enabling computational dynamics in distributed computing environments using a heterogeneous computing template. In: NDIA Ground Vehicle Systems Engineering and Technology Symposium (2011)
59. Negrut, D., Rampalli, R., Ottarsson, G., Sajdak, A.: On an implementation of the Hilber-Hughes-Taylor method in the context of index 3 differential-algebraic equations of multibody dynamics (detc2005-85096). J. Comput. Nonlinear Dyn **2**, 73–85 (2007)
60. Nemat-Nasser, S.: Plasticity: A Treatise on Finite Deformation of Heterogeneous Inelastic Materials. Cambridge University Press, Cambridge (2004)
61. NVIDIA: CUDA Programming Guide (2015). http://docs.nvidia.com/cuda/cuda-c-programming-guide/index.html
62. OpenMP: Specification Standard 4.0 (2013). http://openmp.org/wp/
63. O'Sullivan, C., Bray, J.D.: Selecting a suitable time step for discrete element simulations that use the central difference time integration scheme. Eng. Comput. **21**, 278–303 (2004)
64. Pazouki, A., Negrut, D.: Numerical investigation of microfluidic sorting of microtissues. Comput. Math. Appl. (2015, accepted)
65. Pazouki, A., Negrut, D.: A numerical study of the effect of particle properties on the radial distribution of suspensions in pipe flow. Comput. Fluids **108**, 1–12 (2015)

66. Pazouki, A., Serban, R., Negrut, D.: A high performance computing approach to the simulation of fluid-solid interaction problems with rigid and flexible components. Arch. Mech. Eng. **61**, 227–251 (2014)
67. Pazouki, A., Serban, R., Negrut, D.: A Lagrangian-Lagrangian framework for the simulation of rigid and deformable bodies in fluid. In: Terze, Z. (ed.) Multibody Dynamics. Computaional Methods in Applied Sciences, pp. 33–52. Springer International Publishing, Heidelberg (2014)
68. Pazouki, A., Song, B., Negrut, D.: Boundary condition enforcing methods for smoothed particle hydrodynamics. Technical report: TR-2015-08 (2015)
69. Rankin, C., Nour-Omid, B.: The use of projectors to improve finite element performance. Comput. Struct. **30**, 257–267 (1988)
70. Project Chrono: Chrono: An Open Source Framework for thePhysics-Based Simulation of Dynamic Systems. http://www.projectchrono.org. Accessed 7 Feb 2015
71. Project Chrono: Chrono: An OpenSource Framework for the Physics-Based Simulation of Dynamic Systems. https://github.com/projectchrono/chrono. Accessed 15 Aug 2015
72. Schwertassek, R., Wallrapp, O., Shabana, A.A.: Flexible multibody simulation and choice of shape functions. Nonlinear Dyn. **20**, 361–380 (1999)
73. Serban, R., Mazhar, H., Melanz, D., Jayakumar, P., Negrut, D.: A comparative study of penalty and complementarity methods for handling frictional contact in large multibody dynamics problems. In: 17th U.S. National Congress on Theoretical and Applied Mechanics (USNC-TAM) (2014)
74. Shabana, A.A.: Dynamics of Multibody Systems, 4th edn. Cambridge University Press, Cambridge (2013)
75. Shotwell, R.: A comparison of chrono::engines primitive jointswith ADAMS results. Technical report TR-2012-01, Simulation-Based Engineering Laboratory, University of Wisconsin-Madison (2012). http://sbel.wisc.edu/documents/TR-2012-01.pdf
76. Silbert, L.E., Ertaş, D., Grest, G.S., Halsey, T.C., Levine, D., Plimpton, S.J.: Granular flow down an inclined plane: Bagnold scaling and rheology. Phys. Rev. E **64**, 051302 (2001)
77. Simo, J.C., Rifai, M.: A class of mixed assumed strain methods and the method of incompatible modes. Int. J. Numer. Meth. Eng. **29**, 1595–1638 (1990)
78. Simulation-Based Engineering Lab (SBEL): Movies, Physics-Based Modeling and Simulation. http://sbel.wisc.edu/Animations. Accessed 09 June 2015
79. Simulation-Based Engineering Lab (SBEL): Chrono Vimeo Movies. https://vimeo.com/uwsbel. Accessed 09 June 2015
80. Sin, F.S., Schroeder, D., Barbič, J.: Vega: non-linear FEM deformable object simulator. Comput. Graph. Forum **32**, 36–48 (2013). Wiley Online Library
81. Stewart, D.E.: Rigid-body dynamics with friction and impact. SIAM Rev. **42**(1), 3–39 (2000)
82. Stewart, D.E., Trinkle, J.C.: An implicit time-stepping scheme for rigid-body dynamics with inelastic collisions and Coulomb friction. Int. J. Numer. Meth. Eng. **39**, 2673–2691 (1996)
83. Sugiyama, H., Yamashita, H., Jayakumar, P.: Right on tracks - an integrated tire model for ground vehicle simulation. Tire Technol. Int. **67**, 52–55 (2014)
84. Sugiyama, H., Yamashita, H., Jayakumar, P.: ANCF tire models for multibody ground vehicle simulation. In: Proceedings of International Tyre Colloquium: Tyre Models for Vehicle Dynamics Analysis, 25–28 June 2015
85. Swegle, J., Hicks, D., Attaway, S.: Smoothed particle hydrodynamics stability analysis. J. Comput. Phys. **116**, 123–134 (1995)

86. Tasora, A., Anitescu, M.: A complementarity-based rolling friction model for rigid contacts. Meccanica **48**, 1643–1659 (2013)
87. Tasora, A., Anitescu, M., Negrini, S., Negrut, D.: A compliant visco-plastic particle contact model based on differential variational inequalities. Int. J. Non-Linear Mech. **53**, 2–12 (2013)
88. Taylor, M., Serban, R.: Validation of basic modeling elements in chrono. Technical report TR-2015-05, Simulation-BasedEngineering Laboratory, University of Wisconsin-Madison (2015). http://sbel.wisc.edu/documents/TR-2015-05.pdf
89. Trinkle, J., Pang, J.-S., Sudarsky, S., Lo, G.: On dynamic multi-rigid-body contact problems with Coulomb friction. Zeitschrift fur angewandte Mathematik und Mechanik **77**, 267–279 (1997)
90. Uehara, J.S., Ambroso, M.A., Ojha, R.P., Durian, D.J.: Erratum: low-speed impact craters in loose granular media [phys. rev. lett.prltao0031-9007 90, 194301 (2003)]. Phys. Rev. Lett. **91**, 149902 (2003)
91. Yamashita, H., Matsutani, Y., Sugiyama, H.: Longitudinal tire dynamics model for transient braking analysis: ANCF-LuGre tire model. J. Comput. Nonlinear Dyn. **10**, 031003 (2015)
92. Yamashita, H., Valkeapää, A.I., Jayakumar, P., Sugiyama, H.: Continuum mechanics based bilinear shear deformable shell element using absolute nodal coordinate formulation. J. Comput. Nonlinear Dyn. **10**, 051012 (2015)
93. Zhang, H.P., Makse, H.A.: Jamming transition in emulsions and granular materials. Phys. Rev. E **72**, 011301 (2005)

Parallel Computing in Multi-scale Analysis of Coupled Heat and Moisture Transport in Masonry Structures

Jaroslav Kruis[✉], Tomáš Krejčí, and Michal Šejnoha

Department of Mechanics, Faculty of Civil Engineering,
Czech Technical University in Prague, Thákurova 7, 166 29 Prague, Czech Republic
{jk,krejci,sejnom}@cml.fsv.cvut.cz
http://mech.fsv.cvut.cz/web/

Abstract. This paper describes multi-scale analysis of coupled heat and moisture transport performed on parallel computers. Multi-scale analyses are very computationally demanding because a sequence of problems on low level scales has to be solved many times. The coupled heat and moisture transport is based on the Künzel model and only two scales are used in this contribution. The problems on lower scale are spread among slave processors while the problem on the higher level is solved on the master processor. The multi-scale approach is used for analysis of existing historical masonry structure.

Keywords: Coupled heat and moisture transport · Two-scale approach · Parallel computing

1 Introduction

Parallel computing is used in many scientific and engineering areas at this time. Parallel computers help to solve various real-world problems. This paper describes analysis of coupled heat and moisture transport in highly heterogeneous material, such as masonry, which uses two-scale approach. The motivation for two-scale model comes from the composition of masonry in historical buildings and bridges where irregular stone blocks and layers with variable thickness are encountered. A detailed finite element model of such irregular structures results in millions or billions of unknowns. Problems with such high number of unknowns are hardly solvable but the main obstacle is the randomness of the masonry composition.

Two-scale model offers an advantage of statistical equivalence between the model and the real masonry. There are two levels, the macro-scale level and the meso-scale level. The macro-scale level is used for description of the whole structure and details of the masonry composition are not taken into account while the meso-scale level serves for capturing of details in a statistical sense.

The paper is organized as follows. Section 2 describes very briefly the Künzel model for heat and moisture transport in porous materials. Section 3 is devoted

© Springer International Publishing Switzerland 2016
T. Kozubek et al. (Eds.): HPCSE 2015, LNCS 9611, pp. 50–59, 2016.
DOI: 10.1007/978-3-319-40361-8_3

to the two-scale approach where the basic equations for macro-scale and meso-scale levels are derived. The last Sect. 4 describes parallel implementation and analysis of existing masonry bridge.

2 Künzel Model

Künzel formulated in his thesis [3] in 1995 model which is based on the moisture balance equation in the form

$$\frac{\partial \rho_v}{\partial t} = \text{div} \ (D_w \ \text{grad} \ \rho_v + \delta_p \ \text{grad} \ p_v),\qquad(1)$$

where ρ_v is the partial moisture density (kg/m^3), i.e. the mass of water per unit volume of the porous body, D_w is the capillary water transport coefficient (m^2/s), δ_p is the water vapour permeability (s) and p_v is the partial pressure of water vapour (Pa). Second equation is the heat balance equation in the form

$$\frac{\partial H}{\partial t} = \text{div} \ (\lambda \ \text{grad} \ T) + L_v \ \text{div} \ (\delta_p \ \text{grad} \ p_v),\qquad(2)$$

where H is the enthalpy density (J/m^3), T is the temperature (K), λ is the thermal conductivity (W/m/K), L_v is the latent heat of evaporation of water (J/kg). The relative humidity φ is used for both hygroscopic and overhygroscopic range as the only moisture potential. In the hygroscopic range, standard sorption isotherm $w = w(\varphi)$ is used and w denotes the amount of liquid and gaseous moisture. The balance Eqs. (1) and (2) were modified with the help of relationship

$$p_v = \varphi p_s\qquad(3)$$

between the partial pressure of water vapour, p_v, and the partial pressure of saturated water vapor in the air, p_s, in references [1,2] into new form

$$\frac{\mathrm{d}\rho_v}{\mathrm{d}\varphi}\frac{\partial \varphi}{\partial t} = \text{div} \ \left(\left(D_w \varrho_w \frac{\mathrm{d}w}{\mathrm{d}p_v} + \delta_p p_s\right) \text{grad} \ \varphi + \delta_p \varphi \frac{\mathrm{d}p_s}{\mathrm{d}T} \ \text{grad} \ T\right),\qquad(4)$$

$$\frac{\mathrm{d}H}{\mathrm{d}T}\frac{\partial T}{\partial t} = \text{div} \ \left(\left(\lambda + L_v \delta_p \varphi \frac{\mathrm{d}p_s}{\mathrm{d}T}\right) \text{grad} \ T\right) + \text{div} \ (L_v \delta_p p_s \ \text{grad} \ \varphi),\qquad(5)$$

where ϱ_w is the density of water (kg/m^3). The balance equations are defined on a domain Ω with boundary Γ which is split into three disjoint parts Γ_d, Γ_q and Γ_r where the Dirichlet, Neumann and Newton (Robin) boundary conditions are prescribed.

3 Coupled Heat and Moisture Transport in Masonry

Masonry is highly heterogeneous material where multi-scale methods can be used efficiently. The following subsections describe briefly two-scale method, functions defined on the meso-scale level, balance equations for heat transport, balance equations for moisture transport and the last subsection summarizes the final equations.

3.1 Multi-scale Method

On the macro-scale level, the coordinates are denoted $\boldsymbol{x}$, $T(\boldsymbol{x})$ is the temperature and $\varphi(\boldsymbol{x})$ stands for the relative humidity. All variables on the meso-scale level are denoted by a bar, i.e. $\bar{\boldsymbol{x}}$, $\bar{T}(\boldsymbol{x})$ and $\bar{\varphi}(\boldsymbol{x})$. A function defined on the meso-scale level can be written in the form

$$\bar{f}(\boldsymbol{x}, \bar{\boldsymbol{x}}) = f(\boldsymbol{x}) + (\nabla f(\boldsymbol{x}))^T (\bar{\boldsymbol{x}} - \boldsymbol{x}) + \bar{f}^*(\bar{\boldsymbol{x}}), \tag{6}$$

where $f(\boldsymbol{x})$ is the function defined on the macro-scale level, $\nabla f(\boldsymbol{x})$ is the gradient of the function $f(\boldsymbol{x})$ with respect to the macro-scale level coordinates and $\bar{f}^*(\bar{\boldsymbol{x}})$ is the function of fluctuations defined on the meso-scale level. This formulation corresponds to the first order homogenization where only the function value and the gradient of the macro-scale level function is used on the meso-scale level.

Volume averaging is defined in the form

$$\langle f(\boldsymbol{x}) \rangle = \frac{1}{|\bar{V}|} \int_{\bar{V}} \bar{f}(\boldsymbol{x}, \bar{\boldsymbol{x}}) \mathrm{d}\bar{V} \tag{7}$$

and the surface averaging can be written as

$$\langle f(\boldsymbol{x}) \rangle_S = \frac{1}{|\bar{V}|} \int_{\bar{S}} \bar{f}(\boldsymbol{x}, \bar{\boldsymbol{x}}) \mathrm{d}\bar{S}. \tag{8}$$

The previous averaged values are valid in a macro-scale level point and they are obtained by integration of a function over the representative volume or surface. Volume integral of a function has the form

$$\int_V f(\boldsymbol{x}) \mathrm{d}V = \int_V \langle f(\boldsymbol{x}) \rangle \mathrm{d}V = \int_V \left(\frac{1}{|\bar{V}|} \int_{\bar{V}} \bar{f}(\boldsymbol{x}, \bar{\boldsymbol{x}}) \mathrm{d}\bar{V} \right) \mathrm{d}V. \tag{9}$$

Averaging of coordinates leads to the result

$$\langle \bar{\boldsymbol{x}} - \boldsymbol{x} \rangle = \boldsymbol{0}. \tag{10}$$

In the multi-scale approach, two gradients are defined. One is defined on the macro-scale level and second on the meso-scale level in the form

$$\nabla f(\boldsymbol{x}) = \begin{pmatrix} \dfrac{\partial f}{\partial x_1} \\ \dfrac{\partial f}{\partial x_2} \\ \dfrac{\partial f}{\partial x_3} \end{pmatrix} \qquad \bar{\nabla} \bar{f}(\bar{\boldsymbol{x}}) = \begin{pmatrix} \dfrac{\partial \bar{f}}{\partial \bar{x}_1} \\ \dfrac{\partial \bar{f}}{\partial \bar{x}_2} \\ \dfrac{\partial \bar{f}}{\partial \bar{x}_3} \end{pmatrix}. \tag{11}$$

More details can be found in [4,5] or [6].

3.2   Functions Defined in Meso-Scale Problem

The following functions are needed in the meso-scale problem. The temperature

$$\bar{T}(\boldsymbol{x}, \bar{\boldsymbol{x}}) = T(\boldsymbol{x}) + (\nabla T(\boldsymbol{x}))^T (\bar{\boldsymbol{x}} - \boldsymbol{x}) + \bar{T}^*(\bar{\boldsymbol{x}}), \tag{12}$$

the test function for heat balance equation

$$\bar{\chi}(\boldsymbol{x}, \bar{\boldsymbol{x}}) = \chi(\boldsymbol{x}) + (\nabla \chi(\boldsymbol{x}))^T (\bar{\boldsymbol{x}} - \boldsymbol{x}) + \bar{\chi}^*(\bar{\boldsymbol{x}}), \tag{13}$$

the relative humidity

$$\bar{\varphi}(\boldsymbol{x}, \bar{\boldsymbol{x}}) = \varphi(\boldsymbol{x}) + (\nabla \varphi(\boldsymbol{x}))^T (\bar{\boldsymbol{x}} - \boldsymbol{x}) + \bar{\varphi}^*(\bar{\boldsymbol{x}}), \tag{14}$$

the test function for mass balance equation

$$\bar{\psi}(\boldsymbol{x}, \bar{\boldsymbol{x}}) = \psi(\boldsymbol{x}) + (\nabla \psi(\boldsymbol{x}))^T (\bar{\boldsymbol{x}} - \boldsymbol{x}) + \bar{\psi}^*(\bar{\boldsymbol{x}}). \tag{15}$$

3.3   The Heat Balance Equation

The energy balance equation has to be satisfied on the macro-scale level as well as on the meso-scale level, therefore

$$\frac{\mathrm{d}H}{\mathrm{d}T} \frac{\partial T}{\partial t} = \nabla^T \left(\boldsymbol{D}_{TT} \nabla T + \boldsymbol{D}_{T\varphi} \nabla \varphi \right), \tag{16}$$

$$\frac{\mathrm{d}\bar{H}}{\mathrm{d}\bar{T}} \frac{\partial \bar{T}}{\partial t} = \bar{\nabla}^T \left(\bar{\boldsymbol{D}}_{TT} \bar{\nabla} \bar{T} + \bar{\boldsymbol{D}}_{T\varphi} \bar{\nabla} \bar{\varphi} \right). \tag{17}$$

The matrices $\boldsymbol{D}_{TT}, \boldsymbol{D}_{T\varphi}, \bar{\boldsymbol{D}}_{TT}, \bar{\boldsymbol{D}}_{T\varphi}$ contain material parameters. Equation (16) is multiplied by χ, the balance Eq. (17) is multiplied by the test function $\bar{\chi}$ and then averaged. Finally, both equations are integrated over the volume and their difference leads to the following equation

$$\int_V \left\langle \bar{\chi} \frac{\mathrm{d}\bar{H}}{\mathrm{d}\bar{T}} \frac{\partial \bar{T}}{\partial t} \right\rangle \mathrm{d}V - \int_V \left\langle \bar{\chi} \, \bar{\nabla}^T \left(\bar{\boldsymbol{D}}_{TT} \bar{\nabla} \bar{T} + \bar{\boldsymbol{D}}_{T\varphi} \bar{\nabla} \bar{\varphi} \right) \right\rangle \mathrm{d}V =$$

$$= \int_V \chi \frac{\mathrm{d}H}{\mathrm{d}T} \frac{\partial T}{\partial t} \mathrm{d}V - \int_V \chi \, \nabla^T \left(\boldsymbol{D}_{TT} \nabla T + \boldsymbol{D}_{T\varphi} \nabla \varphi \right) \mathrm{d}V. \tag{18}$$

Substitution of the functions (12), (13), (14) and (15) and the gradients to the heat balance Eq. (18) results in the form

$$\int_V \left\langle (\chi(\boldsymbol{x}) + (\nabla \chi(\boldsymbol{x}))^T (\bar{\boldsymbol{x}} - \boldsymbol{x}) + \bar{\chi}^*(\bar{\boldsymbol{x}})) \frac{\mathrm{d}\bar{H}}{\mathrm{d}\bar{T}} \frac{\partial \bar{T}}{\partial t} \right\rangle \mathrm{d}V +$$

$$+ \int_V \left\langle (\nabla \chi(\boldsymbol{x}) + \bar{\nabla} \bar{\chi}^*(\bar{\boldsymbol{x}}))^T \bar{\boldsymbol{D}}_{TT} (\nabla T(\boldsymbol{x}) + \bar{\nabla} \bar{T}^*(\bar{\boldsymbol{x}})) \right\rangle \mathrm{d}V +$$

$$+ \int_V \left\langle (\nabla \chi(\boldsymbol{x}) + \bar{\nabla} \bar{\chi}^*(\bar{\boldsymbol{x}}))^T \bar{\boldsymbol{D}}_{T\varphi} (\nabla \varphi(\boldsymbol{x}) + \bar{\nabla} \bar{\varphi}^*(\bar{\boldsymbol{x}})) \right\rangle \mathrm{d}V - \tag{19}$$

$$- \int_V \left\langle \bar{\chi} \boldsymbol{n}^T \bar{\boldsymbol{q}}_T \right\rangle_S \mathrm{d}V - \int_V \left\langle \bar{\chi} \boldsymbol{n}^T \bar{\boldsymbol{q}}_\varphi \right\rangle_S \mathrm{d}V = \int_V \chi \frac{\mathrm{d}H}{\mathrm{d}T} \frac{\partial T}{\partial t} \mathrm{d}V +$$

$$\int_V (\nabla \chi)^T \boldsymbol{D}_{TT} \nabla T \mathrm{d}V + \int_V (\nabla \chi)^T \boldsymbol{D}_{T\varphi} \nabla \varphi \mathrm{d}V - \int_S \chi \boldsymbol{n}^T \boldsymbol{q}_T \mathrm{d}S - \int_S \chi \boldsymbol{n}^T \boldsymbol{q}_\varphi \mathrm{d}S.$$

All boundary integrals disappear because zero Neumann boundary conditions are assumed on the macro-scale level for simplicity and with respect to periodicity on the meso-scale level.

The balance equations on both levels are discretized by the finite element method in the classical sense. The number of degrees of freedom on the macro-scale level is denoted N while the number of degrees of freedom on the meso-scale level is denoted n. In the following text, $\boldsymbol{u}_T \in R^N$ is the vector of nodal temperatures, $\boldsymbol{v}_T \in R^N$ is the vector of nodal values of the test function χ, $\boldsymbol{u}_\varphi \in R^N$ is the vector of nodal relative humidities, $\boldsymbol{v}_\varphi \in R^N$ is the vector of nodal values of the test function ψ, $\boldsymbol{B}_T, \boldsymbol{B}_\varphi \in R^{3 \times N}$ are the matrices of gradients of basis functions. The same basis functions are used for a quantity and its test function. All previously mentioned vectors and matrices are defined on the macro-scale level. Similar vectors and matrices $\bar{\boldsymbol{u}}_T^*, \bar{\boldsymbol{v}}_T^*, \bar{\boldsymbol{u}}_\varphi^*, \bar{\boldsymbol{v}}_\varphi^* \in R^n$, $\bar{\boldsymbol{B}}_T, \bar{\boldsymbol{B}}_\varphi \in R^{3 \times n}$ are defined also on the meso-scale level.

Substitution of the finite element approximation to the balance equation on the macro-scale level leads to the form

$$
\int_V \left(\left\langle (\chi(\boldsymbol{x}) + (\nabla\chi(\boldsymbol{x}))^T(\bar{\boldsymbol{x}} - \boldsymbol{x})) \frac{\mathrm{d}\bar{H}}{\mathrm{d}\bar{T}} \frac{\partial \bar{T}}{\partial t} \right\rangle - \chi \frac{\mathrm{d}H}{\mathrm{d}T} \frac{\partial T}{\partial t} \right) \mathrm{d}V
$$
$$
+ \int_V \boldsymbol{v}_T^T \boldsymbol{B}_T^T \left(-\boldsymbol{D}_{TT}\nabla T(\boldsymbol{x}) + \left\langle \bar{\boldsymbol{D}}_{TT}(\nabla T(\boldsymbol{x}) + \bar{\boldsymbol{B}}_T \bar{\boldsymbol{u}}_T^*) \right\rangle \right) \mathrm{d}V \quad (20)
$$
$$
+ \int_V \boldsymbol{v}_T^T \boldsymbol{B}_T^T \left(-\boldsymbol{D}_{T\varphi}\nabla \varphi(\boldsymbol{x}) + \left\langle \bar{\boldsymbol{D}}_{T\varphi}(\nabla \varphi(\boldsymbol{x}) + \bar{\boldsymbol{B}}_\varphi \bar{\boldsymbol{u}}_\varphi^*) \right\rangle \right) \mathrm{d}V = 0.
$$

If new matrices are defined in the form

$$
\boldsymbol{L}_{TT} = \left\langle \bar{\boldsymbol{D}}_{TT} \bar{\boldsymbol{B}}_T \right\rangle \in R^{3 \times n} \qquad \boldsymbol{L}_{T\varphi} = \left\langle \bar{\boldsymbol{D}}_{T\varphi} \bar{\boldsymbol{B}}_\varphi \right\rangle \in R^{3 \times n}, \qquad (21)
$$

the discrete version of the heat balance equation on the macro-scale level is

$$
-\boldsymbol{D}_{TT}\nabla T - \boldsymbol{D}_{T\varphi}\nabla \varphi + \langle \bar{\boldsymbol{D}}_{TT}\rangle \nabla T + \langle \bar{\boldsymbol{D}}_{T\varphi}\rangle \nabla \varphi + \boldsymbol{L}_{TT}\bar{\boldsymbol{u}}_T^* + \boldsymbol{L}_{T\varphi}\bar{\boldsymbol{u}}_\varphi^* = \boldsymbol{0} \quad (22)
$$

and the first integral in (20) is used for definition of the macro-scale capacity matrix. Substitution of the finite element approximation to the balance equation on the meso-scale level leads to the form

$$
\int_V \left\langle (\bar{\boldsymbol{v}}_T^*)^T \bar{\boldsymbol{N}}_T^T \frac{\mathrm{d}\bar{H}}{\mathrm{d}\bar{T}} \bar{\boldsymbol{N}}_T \frac{\mathrm{d}\bar{\boldsymbol{u}}_T^*}{\mathrm{d}t} \right\rangle \mathrm{d}V + \int_V \left\langle (\bar{\boldsymbol{v}}_T^*)^T \bar{\boldsymbol{B}}_T^T \bar{\boldsymbol{D}}_{TT}(\nabla T(\boldsymbol{x}) + \bar{\boldsymbol{B}}_T \bar{\boldsymbol{u}}_T^*) \right\rangle \mathrm{d}V
$$
$$
+ \int_V \left\langle (\bar{\boldsymbol{v}}_T^*)^T \bar{\boldsymbol{B}}_T^T \bar{\boldsymbol{D}}_{T\varphi}(\nabla \varphi(\boldsymbol{x}) + \bar{\boldsymbol{B}}_\varphi \bar{\boldsymbol{u}}_\varphi^*) \right\rangle \mathrm{d}V = 0. \quad (23)
$$

If new matrices are defined in the form

$$
\bar{\boldsymbol{K}}_{TT} = \left\langle \bar{\boldsymbol{B}}_T^T \bar{\boldsymbol{D}}_{TT} \bar{\boldsymbol{B}}_T \right\rangle \in R^{n \times n}, \qquad \bar{\boldsymbol{K}}_{T\varphi} = \left\langle \bar{\boldsymbol{B}}_T^T \bar{\boldsymbol{D}}_{T\varphi} \bar{\boldsymbol{B}}_\varphi \right\rangle \in R^{n \times n},
$$
$$
\bar{\boldsymbol{C}}_{TT} = \left\langle \bar{\boldsymbol{N}}_T^T \frac{\mathrm{d}\bar{H}}{\mathrm{d}\bar{T}} \bar{\boldsymbol{N}}_T \right\rangle \in R^{n \times n},
$$

the discrete version of the heat balance equation on the meso-scale level is

$$\bar{C}_{TT}\frac{d\bar{u}_T^*}{dt} + L_{TT}^T\nabla T(x) + L_{T\varphi}^T\nabla\varphi(x) + \bar{K}_{TT}\bar{u}_T^* + \bar{K}_{T\varphi}\bar{u}_\varphi^* = 0. \quad (24)$$

3.4 The Mass Balance Equation

Discretized version of the mass balance equations on the macro and meso-scale levels can be obtained in similar way to the equations for heat transport. The whole derivation is not included and only the resulting equations are summarized. The mass balance equation on the macro-scale level has the form

$$- D_{\varphi T}\nabla T - D_{\varphi\varphi}\nabla\varphi + \langle\bar{D}_{\varphi T}\rangle\nabla T + \langle\bar{D}_{\varphi\varphi}\rangle\nabla\varphi + L_{\varphi T}\bar{u}_T^* + L_{\varphi\varphi}\bar{u}_\varphi^* = 0, \quad (25)$$

while the mass balance equation on the meso-scale level has the form

$$\bar{C}_{\varphi\varphi}\frac{d\bar{u}_\varphi^*}{dt} + L_{\varphi T}^T\nabla T(x) + L_{\varphi\varphi}^T\nabla\varphi(x) + \bar{K}_{\varphi T}\bar{u}_T^* + \bar{K}_{\varphi\varphi}\bar{u}_\varphi^* = 0. \quad (26)$$

3.5 Solution of Multi-scale Problem

The multi-scale heat and moisture transport is described by two balance equations on the macro-scale level and two equations on the meso-scale level. On the meso-scale level, the balance equations have the form

$$\bar{C}_{TT}\frac{d\bar{u}_T^*}{dt} + L_{TT}^T\nabla T(x) + L_{T\varphi}^T\nabla\varphi(x) + \bar{K}_{TT}\bar{u}_T^* + \bar{K}_{T\varphi}\bar{u}_\varphi^* = 0, \quad (27)$$

$$\bar{C}_{\varphi\varphi}\frac{d\bar{u}_\varphi^*}{dt} + L_{\varphi T}^T\nabla T(x) + L_{\varphi\varphi}^T\nabla\varphi(x) + \bar{K}_{\varphi T}\bar{u}_T^* + \bar{K}_{\varphi\varphi}\bar{u}_\varphi^* = 0. \quad (28)$$

On the macro-scale level, the relationship for the conductivity matrices has the form

$$- D_{TT}\nabla T - D_{T\varphi}\nabla\varphi + \langle\bar{D}_{TT}\rangle\nabla T + \langle\bar{D}_{T\varphi}\rangle\nabla\varphi + L_{TT}\bar{u}_T^* + L_{T\varphi}\bar{u}_\varphi^* = 0, \quad (29)$$

$$-D_{\varphi T}\nabla T - D_{\varphi\varphi}\nabla\varphi + \langle\bar{D}_{\varphi T}\rangle\nabla T + \langle\bar{D}_{\varphi\varphi}\rangle\nabla\varphi + L_{\varphi T}\bar{u}_T^* + L_{\varphi\varphi}\bar{u}_\varphi^* = 0, \quad (30)$$

which can be rearranged to the form

$$\begin{pmatrix} D_{TT} & D_{T\varphi} \\ D_{\varphi T} & D_{\varphi\varphi} \end{pmatrix} = \begin{pmatrix} \langle\bar{D}_{TT}\rangle & \langle\bar{D}_{T\varphi}\rangle \\ \langle\bar{D}_{\varphi T}\rangle & \langle\bar{D}_{\varphi\varphi}\rangle \end{pmatrix} - \quad (31)$$

$$- \begin{pmatrix} L_{TT} & L_{T\varphi} \\ L_{\varphi T} & L_{\varphi\varphi} \end{pmatrix} \begin{pmatrix} \bar{K}_{TT} & \bar{K}_{T\varphi} \\ \bar{K}_{\varphi T} & \bar{K}_{\varphi\varphi} \end{pmatrix}^{-1} \begin{pmatrix} L_{TT}^T & L_{T\varphi}^T \\ L_{\varphi T}^T & L_{\varphi\varphi}^T \end{pmatrix}.$$

The conductivity matrix of the material (31) is used on the macro-scale level and it is obtained by homogenization from the meso-scale level. The conductivity matrix of the macro-scale problem has the form

$$K = \int_V \begin{pmatrix} B_T \\ B_\varphi \end{pmatrix}^T \begin{pmatrix} D_{TT} & D_{T\varphi} \\ D_{\varphi T} & D_{\varphi\varphi} \end{pmatrix} \begin{pmatrix} B_T \\ B_\varphi \end{pmatrix} dV \quad (32)$$

Additional details can be found in references [5,6].

4   Two-Scale Analysis of Heat and Moisture Transport in Masonry Bridge

As was mentioned earlier, there are two levels. The macro-scale level describes the whole structure and it does not take into account the particular composition of the masonry. Material parameters have to be obtained from Eq. (31). It means, in every integration point of the macro-scale problem a meso-scale problem described by the system of Eqs. (27) and (28) has to be solved. Clearly, it is very computationally demanding and it is suitable to use a parallel computer.

In our implementation, the macro-scale problem is assigned to the master processor while all meso-scale problems are spread among the slave processors. We analyzed distribution of the temperature and relative humidity in Charles bridge in Prague, Czech Republic. The heat and moisture transport was solved with the help of two and three-dimensional models.

We used two-dimensional model first. The finite element mesh used at the macro-level is evident in Fig. 1 featuring 7,081 nodes and 13,794 triangular elements with a single integration point thus amounting to the solution of 13,794 meso-problems at each macroscopic time step. This figure also shows decomposition of the macro-problem into 12 slave processors. It should be noted that the assumed decomposition of the macro-problem is not ideal. In comparison with domain decomposition methods, the macro-problem has to be split with respect to the heterogeneity of the material resulting in the variation of number of elements in individual sub-domains between 1046 and 1748. The actual analysis was performed on a cluster built at our department. Each node of the cluster is a single processor personal computer Dell Optiplex GX620 equipped with 3.54 GB of RAM. The processors are Intel Pentium with the frequency 3.4 GHz. The cluster is based on Debian linux 5.0 and 32-bit architecture. Each time step in the macro-level problem took 2.08 min.

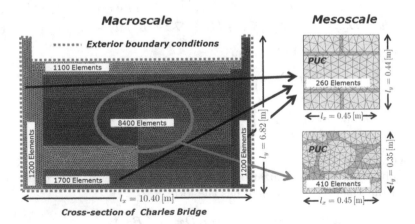

Fig. 1. Two-dimensional model: cross section of the bridge and two representative cells.

When the two-dimensional analysis was finished, we extended the model from two to three dimensions. The three-dimensional numerical model simulates the coupled heat and moisture transfer of one half of arch III of the bridge. The finite element mesh of the macro-problem was created using tetrahedron elements with linear approximation functions. It contains 73,749 nodes and 387,773 elements and it is visible in Fig. 2. Meso-problem is assigned to each finite element of the macro-problem. The mesh of macro-problem was decomposed into 15 sub-domains and the appropriate meso-problems were aggregated and assigned to 15 slave processors. Regular masonry (arch vault, pier and breast walls) is represented by SEPUC (statistically equivalent periodic unit cell) No. 1 which is formed from 1950 nodes and 1512 hexahedron elements with linear approximation functions, while irregular masonry (infill) is represented by SEPUC No. 2 with 2454 nodes and 12269 tetrahedron elements with linear approximation functions. Both cells are in Fig. 3. Other parts of the bridge (concrete slab and layers of pavement) are considered as homogeneous and isotropic. Figure 4 shows distribution of temperature in the bridge and Fig. 5 contains comparison between temperature obtained from numerical analysis and measured temperature.

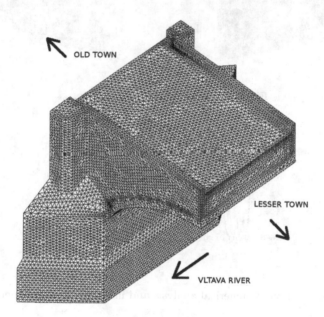

Fig. 2. Macro-scale level: finite element mesh.

The three-dimensional analysis of Charles bridge was performed on DELL computer Precision T5600 equipped with two INTEL XEON processors E5 (2X8 cores) 2.4 GHz and RAM 16 GB (4X4 GB) 1600 MHZ DDR3. The computer is based on Debian linux and 64-bit architecture. The computational time of determination of material properties on meso-structural levels and one time step on macro-level for the problem took 22 h. Decomposition of the macro-scale problem

Fig. 3. Periodic unit cells for regular and irregular masonry.

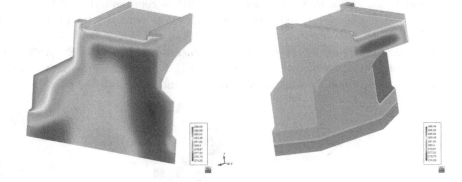

Fig. 4. Distribution of temperature.

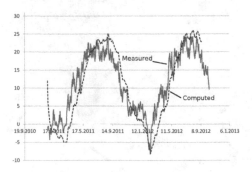

Fig. 5. Comparison of numerical analysis and measured data (temperature).

into 15 subdomains with respect to appropriate type of SEPUC and distribution of the computational requirements among 15 slave processors resulted in significant time reduction. One time step in the macro-scale problem took 10 min.

5   Conclusions

Two-scale analysis of coupled heat and moisture transport based on the Künzel model was described. Governing equations for the macro-scale and meso-scale problems were summarized. The two-scale analysis was implemented in parallel and it was used for solution of existing masonry structure. The macro-scale problems were assigned to the master processor and the meso-scale problems were dedicated to the slave processors. Good scalability was observed because limited amount of data is sent among processors. In fact, any number of slave processors can be used in this type of computation and very good speedup should be obtained.

Acknowledgement. Financial support for this work was provided by project number 15-17615S of Czech Science Foundation. The financial support is gratefully acknowledged.

References

1. Černý, R., Maděra, J., Kočí, J., Vejmelková, E.: Heat and moisture transport in porous materials involving cyclic wetting and drying. In: Proceedings of the Fourteenth International Conference on Computational Methods and Experimental Measurements, pp. 3–12. WIT Press, Wessex, UK (2009)
2. Kočí, V., Maděra, J., Černý, R.: Computer aided design of interior thermal insulation system suitable for autoclaved aerated concrete structures. Appl. Therm. Eng. **58**, 165–172 (2013)
3. Künzel, H.M.: Simultaneous heat and moisture transport in building components. Fraunhofer Institute of Building Physics (1995)
4. Larsson, F., Runesson, K., Su, F.: Variationally consistent computational homogenization of transient heat flow. Int. J. Numer. Meth. Eng. **81**, 1659–1686 (2010)
5. Sýkora, J., Krejčí, T., Kruis, J., Šejnoha, M.: Computational homogenization of non-stationary transport processes in masonry structures. J. Comput. Appl. Math. **236**(18), 4745–4755 (2012)
6. Sýkora, J., Vorel, J., Krejčí, T., Šejnoha, M., Šejnoha, J.: Analysis of coupled heat and moisture transfer in masonry structures. Mater. Struct. **42**(8), 1153–1167 (2009)

An Equation Error Approach for the Identification of Elastic Parameters in Beams and Plates with H_1 Regularization

P. Caya[1], B. Jadamba[1], A.A. Khan[1(✉)], F. Raciti[2], and B. Winkler[3]

[1] Center for Applied and Computational Mathematics,
School of Mathematical Sciences, Rochester Institute of Technology,
85 Lomb Memorial Drive, Rochester, NY 14623, USA
{poc1673,bxjsma,aaksma}@rit.edu

[2] Department of Mathematics and Computer Science, University of Catania,
95125 Catania, Italy
fraciti@dmi.unict.it

[3] Institute of Mathematics, Martin-Luther-University Halle-Wittenberg,
Halle (Saale), Germany
brian.winkler@mathematik.uni-halle.de

Abstract. In this short note deals with the nonlinear inverse problem of identifying a variable parameter in fourth-order partial differential equations using an equation error approach. These equations arise in several important applications such as car windscreen modeling, deformation of plates, etc. To counter the highly ill-posed nature of the considered inverse problem, a regularization must be performed. The main contribution of this work is to show that the equation error approach permits the use of H^1 regularization whereas other optimization-based formulations commonly use H_2 regularization. We give the existence and convergence results for the equation error formulation. An illustrative numerical example is given to show the feasibility of the approach.

Keywords: Inverse problem · Equation error method · Fourth-order boundary value problem · Regularization · Parameter identification

1 Introduction

Let Ω be a bounded open domain in R^2 with a sufficiently smooth boundary Γ and let $f \in L^2(\Omega)$ be a given function. Consider the following fourth-order elliptic boundary value problem

$$\Delta(a\Delta u) = f \quad \text{in} \quad \Omega, \tag{1}$$

augmented with the clamped boundary conditions,

$$u = 0 \quad \text{on} \quad \Gamma, \tag{2a}$$

$$\frac{\partial u}{\partial n} = 0 \quad \text{on} \quad \Gamma. \tag{2b}$$

Dedicated to Prof. Alemdar Hasanoglu (Hasanov) on his 60th birthday

© Springer International Publishing Switzerland 2016
T. Kozubek et al. (Eds.): HPCSE 2015, LNCS 9611, pp. 60–67, 2016.
DOI: 10.1007/978-3-319-40361-8_4

In this work, our objective is to study the inverse problem of identifying the material parameter a from a measurement z of u. Applications of this study are in beam and plate models as well as car windshield modeling (see [16,17]). This nonlinear inverse problem has been explored using the output least squares (OLS) approach in which one attempts to find a minimizer of the functional

$$J(a) := \frac{1}{2}\|u(a) - z\|^2,$$

defined by using a suitable norm (see White [18]). Here z is the data (a measurement of u) and $u(a)$ is the unique solution of (1) that corresponds to the material parameter a,

One of the primary obstacles in a satisfactory treatment of the OLS-based optimization framework is due to the fact that the OLS, in general, is nonconvex. Our objective then is to investigate an equation error approach for solving the nonlinear inverse problem of identifying the material parameter a. In contrast to the OLS based optimization approach, the equation error approach results in solving a convex optimization problem. Some recent developments in parameter identification problems can be found in [2,4–7,7–10,13,14] and the cited references therein. Very interesting study of an identification problem in more general plate models can be found in Hasanov and Mamedov [12], see also [11].

We emphasize that the equation error approach has two advantages over the OLS approach. Firstly, it leads to a convex optimization problem and hence it only possesses global solutions. Secondly, the equation approach is computationally quite inexpensive as there is no underlying variational problem to be solved. On the other hand, a deficiency of the approach is that, due to the fact that it relies on differentiating the data, it is quite sensitive to data contamination.

The equation error approach has been studied in the context of the following simpler second-order elliptic boundary valued problem:

$$-\nabla \cdot (a\nabla u) = f \quad \text{in} \quad \Omega, \tag{3a}$$
$$u = 0 \quad \text{on} \quad \Gamma. \tag{3b}$$

For (3), the equation error approach consists of finding a minimizer of the functional

$$a \to \frac{1}{2}\|\nabla \cdot (a\nabla z) + f\|^2_{H^{-1}(\Omega)},$$

where $H^{-1}(\Omega)$ is the topological dual of $H^1_0(\Omega)$ and z is again the measured data.

In this paper, we extend the equation error approach to identify the coefficient a in the fourth-order boundary value problem (1). Our strategy is motivated by the ideas presented originally by Acar [1] and Kärkkäinen [15] for (3) (see also [3]). Besides giving an existence theorem and a convergence result for the discretized problem, we also some numerical examples.

This paper is divided into four main sections. Section 2 provides essential background material for the problem and poses the solution of the inverse problem as a solvable minimization problem. Section 3 examines the stability of the equation error method and Sect. 4 provides a brief numerical example to show the preliminary computational feasibility of the proposed method.

2   Equation Error Approach

The variational formulation of (1) will be instrumental in formulating the equation error approach. The space suitable for the variational formulation is given by

$$V := \{v \in H^2(\Omega) : \quad u = \frac{\partial u}{\partial n} = 0 \text{ on } \Gamma\}.$$

By multiplying (1) by a test function $v \in V$ and repeatedly using the well-known Green's formula we obtain the following variational formulation of (1): Find $u \in V$ such that

$$\int_\Omega a \Delta u \, \Delta v = \int_\Omega f v, \quad \text{for every } v \in V. \tag{4}$$

For a fixed pair $(a, w) \in L^\infty(\Omega) \times V$, we define the maps $E(a, w) : V \to V^*$ and $m : V \to R$ by

$$E(a, w)(v) = \int_\Omega a \Delta w \, \Delta v,$$

$$m(v) = \int_\Omega f v.$$

We note that, although the functional $E(a, w)$ was defined for fixed $a \in L^\infty(\Omega)$, $w \in V$, it remains well-defined for $a \in L^2(\Omega)$ and $w \in V \cap W^{2,\infty} := V^\infty$. In other words, we can sacrifice some regularity in a by requiring more regularity of u. This fact will play an important role below.

We first prove the following technical result for later use.

Lemma 1. *Assume that $u \in V^\infty$, $a \in L^2(\Omega)$, and $\{a_n\} \subset L^2(\Omega)$ is a sequence such that $a_n \to a$ in $L^2(\Omega)$. Then $E(a_n, u) \to E(a, u)$ in V^*.*

Proof. We begin by showing that the following inequality holds:

$$\|E(a, u)\|_{V^*} \le \|a\|_{L^2} \|u\|_{V^\infty}. \tag{5}$$

In fact, using the definition of E, we have

$$|E(a, u)(v)| \le \left| \int_\Omega a \Delta u \, \Delta v \right| \le \|a \Delta u\|_{L^2} \|\Delta v\|_{L^2},$$

where

$$\|a \Delta u\|_{L^2}^2 = \int_\Omega a^2 (\Delta u)^2 \le \|u\|_{V^\infty}^2 \|a\|_{L^2}^2,$$

and because $\|\Delta v\|_{L^2} \le \|v\|_V$, we at once obtain (5).

To prove the main argument, we note that

$$(E(a_n, u) - E(a, u))(v) = \int_\Omega a_n \Delta u \Delta v - \int_\Omega a \Delta u \Delta v = \int_\Omega (a_n - a) \Delta u \Delta v,$$

which by using (5) implies that

$$|(E(a_n, u) - E(a, u))(v)| \leq \|u\|_{V^\infty} \|a_n - a\|_{L^2} \|v\|_V,$$

and consequently $\|E(a_n, u) - E(a, u)\|_{V^*} \leq \|u\|_{V^\infty} \|a_n - a\|_{L^2}$. The proof is complete. $\square$

Since the inverse problem at hand is ill-posed, some regularization is necessary. For this, we first define $A \subset H^1(\Omega)$ to be the closed and convex set of admissible coefficients. We consider the following regularized equation error functional to estimate a^* from a measurement z of u^* by minimizing

$$J(a; z, \varepsilon) = \|E(a, z) - m\|_{V^*}^2 + \varepsilon \|a\|_{H^1}^2. \tag{6}$$

Here it is assumed that $a^* \in A$ and $u^* \in V$ satisfy (1), $\varepsilon > 0$ is a regularizing parameter, $z \in V$ is the data, and $\| \cdot \|_2^2$ is the regularization term.

Assuming that the data z is sufficiently smooth, we show that $J(\cdot; z, \varepsilon)$ has a unique minimizer in $H^1(\Omega)$ for each $\varepsilon > 0$.

Theorem 1. *Suppose $z \in W^\infty$. Then, for each $\varepsilon > 0$, there exists a unique a_ε satisfying*

$$J(a_\varepsilon; z, \varepsilon) \leq J(a; z, \varepsilon), \text{ for all } a \in H^1(\Omega).$$

Proof. Since the functional J is bounded below, there exists a minimizing sequence $\{a_n\}$ for J. We have $\varepsilon \|a_n\|_{H^1}^2 \leq J(a_n; z, \varepsilon)$ for all n which implies that $\{a_n\}$ is bounded in $H^1(\Omega)$. Therefore, there exists $a_\varepsilon \in H^1(\Omega)$ and a subsequence of $\{a_n\}$ (still denoted by $\{a_n\}$) such that $a_n \to a_\varepsilon$ weakly in $H^1(\Omega)$ and, by Rellich's theorem, strongly in $L^2(\Omega)$. Since $z \in V^\infty$ and $a_n \to a_\varepsilon$ in $L^2(\Omega)$, Lemma 1 confirms that $E(a_n, z) \to E(a_\varepsilon, z)$ and since the norm is weakly lower semicontinuous, it follows that

$$\inf_{a \in H^1(\Omega)} J(a; z, \varepsilon) = \lim_{n \to \infty} J(a_n; z, \varepsilon)$$

$$= \lim_{n \to \infty} \left(\|E(a_n, z) - m\|_{V^*}^2 + \varepsilon \|a_n\|_{H^1}^2 \right)$$

$$\geq \|E(a_\varepsilon, z) - m\|_{V^*}^2 + \varepsilon \|a_\varepsilon\|_{H^1}^2$$

$$= J(a_\varepsilon; z, \varepsilon),$$

confirming that a_ε is a minimizer of $J(\cdot; z, \varepsilon)$. The uniqueness of a_ε follows from the fact that the regularized equation error functional is strictly convex. The proof is complete. $\square$

Since $J(a_\varepsilon; z, \varepsilon) \geq \inf_{a \in H^1(\Omega)} J(a; z, \varepsilon)$, the last inequality in the above proof must actually hold as an equality and hence $\lim_{n \to \infty} \|a_n\|_{H^1} = \|a_\varepsilon\|_{H^1}$ must remain valid. This, in view of the fact $a_n \to a_\varepsilon$ weakly in $H^1(\Omega)$, ensures that $\{a_n\}$ actually converges to a_ε strongly in $H^1(\Omega)$. Consequently any minimizing sequence of $J(\cdot; z, \varepsilon)$ converges in $H^1(\Omega)$ to the unique minimizer a_ε of $J(\cdot; z, \varepsilon)$.

3   Stability of the Equation Error Method

Recall that $a^* \in A$ and $u^* \in V$ are assumed to satisfy (1). However, since a^* is not unique, we define the convex set $S = \{a \in H^1(\Omega) : E(a, u^*) = m\}$.

We can now prove the following stability result for the equation error approach.

Theorem 2. *Assume that $u^* \in V^\infty$ and $a^* \in H^1(\Omega)$ satisfy (1). Let $\{z_n\} \subset V^\infty$ be a sequence of observations of u^* that satisfy, with the sequences $\{\delta_n\}$, $\{\varepsilon_n\}$, the conditions*

1. $\delta_n^2 \leq \varepsilon_n \leq \delta_n$ for all $n \in \mathbb{Z}^+$;
2. $\delta_n^2 / \varepsilon_n \to 0$ as $n \to \infty$;
3. $\|z_n - u^*\|_{V^\infty} \leq \delta_n$ for all $n \in \mathbb{Z}^+$;
4. $\delta_n \to 0$ as $n \to \infty$.

For each $n \in \mathbb{Z}^+$, let a_n be the unique solution of

$$\min_{a \in H^1(\Omega)} J(a; z_n, \varepsilon_n).$$

Then, there exists $\tilde{a} \in S$ such that $a_n \to \tilde{a}$ in $H^1(\Omega)$. Moreover, a satisfies $\|\tilde{a}\|_{H^1} \leq \|a\|_{H^1}$, for all $a \in S$.

Proof. Let $a \in S$ be arbitrary. Then,

$$\varepsilon_n \|a_n\|_{H^1}^2 \leq \|E(a, z_n) - m\|_{V^*}^2 + \varepsilon_n \|a\|_{H^1}^2$$
$$= \|E(a, z_n - u^*)\|_{V^*}^2 + \varepsilon_n \|a\|_{H^1}^2$$
$$\leq c\|a\|_{L^2}^2 \|z_n - u^*\|_{V^\infty}^2 + \varepsilon_n \|a\|_{H^1}^2,$$

implying that

$$\|a_n\|_{H^1}^2 \leq \|a\|_{L^2}^2 \frac{\delta_n^2}{\varepsilon_n} + \|a\|_{H^1}^2, \tag{7}$$

and, in particular,

$$\|a_n\|_{H^1}^2 \leq \|a^*\|_{L^2}^2 \frac{\delta_n^2}{\varepsilon_n} + \|a^*\|_{H^1}^2 \leq \|a^*\|_{L^2}^2 + \|a^*\|_{H^1}^2,$$

where we used the assumption $\delta_n^2 \leq \varepsilon_n$. This proves that $\{a_n\}$ is bounded in $H^1(\Omega)$. Hence, by Rellich's lemma, there exists $\tilde{a} \in H^1(\Omega)$ and a subsequence $\{a_{n_k}\}$ such that $a_{n_k} \to \tilde{a}$ weakly in $H^1(\Omega)$ and strongly in $L^2(\Omega)$.

We claim that $\tilde{a} \in S$. Indeed, for any $\hat{a} \in S$, we have

$$\|E(a_{n_k}, u^*) - m\|_{V^*}^2 = \|E(a_{n_k}, u^*) - E(a_{n_k}, z_{n_k}) + E(a_{n_k}, z_{n_k}) - m\|_{V^*}^2$$
$$\leq 2\|E(a_{n_k}, u^* - z_{n_k})\|_{V^*}^2 + 2\|E(a_{n_k}, z_{n_k}) - m\|_{V^*}^2$$
$$\leq 2\|a_{n_k}\|_{L^2}^2 \|z_{n_k} - u^*\|_{V^\infty}^2 + 2\|E(\hat{a}, z_{n_k}) - m\|_{V^*}^2 + 2\varepsilon_{n_k} \|\hat{a}\|_{H^1}^2$$
$$\leq 2\|a_{n_k}\|_{L^2}^2 \delta_{n_k}^2 + 2\|\hat{a}\|_{L^2}^2 \delta_{n_k}^2 + 2\varepsilon_{n_k} \|\hat{a}\|_{H^1}^2$$
$$\leq 2\|a_{n_k}\|_{L^2}^2 \delta_{n_k}^2 + 4\|\hat{a}\|_{H^1}^2 \delta_{n_k},$$

where we used $\delta_{n_k}^2 \leq \varepsilon_{n_k} \leq \delta_{n_k}$ and the following inequality which remains true for any $\hat{a} \in S$:

$$\|E(\hat{a}, z_{n_k}) - m\|_{V^*}^2 + \varepsilon_{n_k}\|\hat{a}\|_{H^1}^2 \leq \|\hat{a}\|_{L^2}^2 \delta_{n_k}^2 + \varepsilon_{n_k}\|\hat{a}\|_{H^1}^2.$$

Because $\{\|a_{n_k}\|_{L^2}\}$ is bounded and $\delta_{n_k} \to 0$ as $k \to \infty$, this ensures that $\|E(a_{n_k}, u^*) - m\|_{V^*} \to 0$. Since we also have $E(a_{n_k}, u^*) \to E(\tilde{a}, u^*)$ by Lemma 1, this shows that $E(\tilde{a}, u^*) = m$ and hence that $\tilde{a} \in S$.

Using the fact that $a_{n_k} \to \tilde{a}$ weakly in $H^1(\Omega)$, we have $\|\tilde{a}\|_{H^1} \leq \liminf_{k\to\infty} \|a_{n_k}\|_{H^1}$. Moreover, by (7),

$$\varepsilon_{n_k}\|a_{n_k}\|_{H^1}^2 \leq \|\tilde{a}\|_{L^2}^2 \delta_{n_k}^2 + \varepsilon_{n_k}\|\tilde{a}\|_{H^1}^2,$$

which implies that

$$\|a_{n_k}\|_{H^1}^2 \leq \|\tilde{a}\|_{L^2}^2 \frac{\delta_{n_k}^2}{\varepsilon_{n_k}} + \|\tilde{a}\|_{H^1}^2.$$

Since $\delta_{n_k}^2/\varepsilon_{n_k} \to 0$ as $k \to \infty$, this shows that $\limsup_{k\to\infty} \|a_{n_k}\|_{H^1} \leq \|\tilde{a}\|_{H^1}$. Therefore,

$$\|\tilde{a}\|_{H^1} \leq \liminf_{k\to\infty} \|a_{n_k}\|_{H^1} \leq \limsup_{k\to\infty} \|a_n\|_{H^1} \leq \|\tilde{a}\|_{H^1},$$

which shows that $\|a_{n_k}\|_{H^1} \to \|\tilde{a}\|_{H^1}$, and hence that $a_{n_k} \to a$ strongly in $H^1(\Omega)$ as $k \to \infty$.

Using (7),

$$\|\tilde{a}\|_{H^1}^2 \leq \lim_{k\to\infty} \|a_{n_k}\|_{H^1}^2 \leq \lim_{k\to\infty} \left(\|a\|_{L^2}^2 \frac{\delta_{n_k}^2}{\varepsilon_{n_k}} + \|a\|_{H^1}^2 \right) = \|a\|_{H^1}^2$$

holds for every $a \in S$.

Finally, since the set S is a convex, there is a unique minimal H^1-norm element, and we have shown that every convergent subsequence of $\{a_n\}$ converges to this unique element $\tilde{a}$. Thus the whole sequence $\{a_n\}$ must converge to $\tilde{a}$. This completes the proof.   □

4   Numerical Results

To test the preliminary effectiveness of the equation error approach for this inverse problem, we consider an example boundary value problem derived from (1):

$$\Delta\left[a(x,y)\Delta u(x,y)\right] = f(x,y) \qquad \text{in } \Omega$$
$$u(x,y) = \frac{\partial u}{\partial n} = 0 \qquad \text{on } \Gamma \tag{8}$$

where the solution u and parameter a are defined as

$$u(x,y) = 16x^2(1-x)^2 y^2(1-y)^2,$$
$$a(x,y) = 4 + \sin(2\pi x)\sin(3\pi y).$$

For means of this numerical experiment, we take $f(x, y)$ as subsequently defined by (8). The domain Ω is taken as the unit square, $\Omega = (0, 1) \times (0, 1)$ with the boundary Γ as the square's outside edges.

Discretization of the solution was performed using cubic Hermite finite elements on a 20×20 mesh consisting of 882 triangles and 2,048 degrees of freedom.

The discretized optimization problem was solved using a conjugate-gradient trust-region method (`cgtrust`) with a stopping criteria on $\|\nabla J\|$ of 10^{-12}. Using a value of $\varepsilon = 10^{-6}$ for the regularization parameter with the H^1-norm, `cgtrust` converged in 38 iterations. The computed solution at several iterations of the algorithm along with the output of the optimization are summarized in Fig. 1. We note that this method provides a good reconstruction of the parameter in the interior of Ω with reconstruction error concentrated mostly along the boundaries.

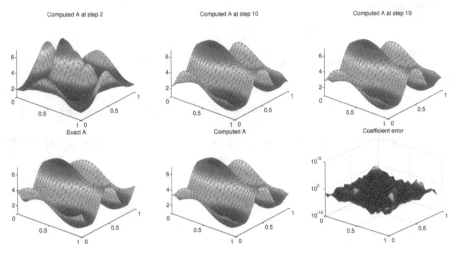

Fig. 1. Parameter recovery using the EE method and `cgtrust` with $\varepsilon = 10^{-6}$.

Acknowledgments. The work of A.A. Khan is supported by a grant from the Simons Foundation (#210443 to Akhtar Khan). The work of B. Jadamba is supported by RITs COS Faculty Development Grant (FEAD) for 2014.

References

1. Acar, R.: Identification of the coefficient in elliptic equations. SIAM J. Control Optim. **31**(5), 1221–1244 (1993)
2. Al-Jamal, M.F., Gockenbach, M.S.: Stability and error estimates for an equation error method for elliptic equations. Inverse Prob. 28(9), 095006-15 (2012)
3. Radu, V.: Application. In: Radu, V. (ed.) Stochastic Modeling of Thermal Fatigue Crack Growth. ACM, vol. 1, pp. 63–70. Springer, Heidelberg (2015)
4. Crossen, E., Gockenbach, M.S., Jadamba, B., Khan, A.A., Winkler, B.: An equation error approach for the elasticity imaging inverse problem for predicting tumor location. Comput. Math. Appl. **67**(1), 122–135 (2014)

5. Gockenbach, M.S.: The output least-squares approach to estimating Lamé moduli. Inverse Prob. **23**(6), 2437–2455 (2007)
6. Gockenbach, M.S., Jadamba, B., Khan, A.A.: Numerical estimation of discontinuous coefficients by the method of equation error. Int. J. Math. Comput. Sci. **1**(3), 343–359 (2006)
7. Gockenbach, M.S., Jadamba, B., Khan, A.A.: Equation error approach for elliptic inverse problems with an application to the identification of Lamé parameters. Inverse Probl. Sci. Eng. **16**(3), 349–367 (2008)
8. Gockenbach, M.S., Khan, A.A.: Identification of Lamé parameters in linear elasticity: a fixed point approach. J. Ind. Manag. Optim. **1**(4), 487–497 (2005)
9. Gockenbach, M.S., Khan, A.A.: An abstract framework for elliptic inverse problems. I. An output least-squares approach. Math. Mech. Solids **12**(3), 259–276 (2007)
10. Gockenbach, M.S., Khan, A.A.: An abstract framework for elliptic inverse problems. II. An augmented Lagrangian approach. Math. Mech. Solids **14**(6), 517–539 (2009)
11. Hasanov, A.: Variational approach to non-linear boundary value problems for elasto-plastic incompressible bending plate. Int. J. Non-Linear Mech. **42**(5), 711–721 (2007). http://dx.doi.org/10.1016/j.ijnonlinmec.2007.02.011
12. Hasanov, A., Mamedov, A.: An inverse problem related to the determination of elastoplastic properties of a plate. Inverse Prob. **10**(3), 601–615 (1994). http://stacks.iop.org/0266-5611/10/601
13. Jadamba, B., Khan, A.A., Rus, G., Sama, M., Winkler, B.: A new convex inversion framework for parameter identification in saddle point problems with an application to the elasticity imaging inverse problem of predicting tumor location. SIAM J. Appl. Math. **74**(5), 1486–1510 (2014)
14. Jadamba, B., Khan, A.A., Sama, M.: Inverse problems of parameter identification in partial differential equations. In: Mathematics in Science and Technology, World Scientific Publishing, Hackensack, NJ, 2011, pp. 228–258
15. Kärkkäinen, T.: An equation error method to recover diffusion from the distributed observation. Inverse Prob. **13**(4), 1033–1051 (1997)
16. Kügler, P.: A parameter identification problem of mixed type related to the manufacture of car windshields. SIAM J. Appl. Math. **64**(3), 858–877 (2004). (electronic)
17. White, L.W.: Estimation of elastic parameters in beams and certain plates: H^1 regularization. J. Optim. Theory Appl. **60**(2), 305–326 (1989)
18. White, L.W.: Estimation of elastic parameters in a nonlinear elliptic model of a plate. Appl. Math. Comput. **42**(2, part II), 139–187 (1991)

A Comparison of Preconditioning Methods for Saddle Point Problems with an Application to Porous Media Flow Problems

Owe Axelsson, Radim Blaheta, and Martin Hasal[✉]

Institute of Geonics AS CR, Studentska 1768,
70800 Ostrava-Poruba, Czech Republic
martin.hasal@ugn.cas.cz

Abstract. The paper overviews and compares some block preconditioners for the solution of saddle point systems, especially systems arising from the Brinkman model of porous media flow. The considered preconditioners involve different Schur complements as inverse free Schur complement in HSS (Hermitian - Skew Hermitian Splitting preconditioner), Schur complement to the velocity matrix and finally Schur complement to a regularization block in the augmented matrix preconditioner. The inverses appearing in most of the considered Schur complements are approximated by simple sparse approximation techniques as element-by-element and Frobenius norm minimization approaches. A special interest is devoted to problems involving various Darcy, Stokes and Brinkman flow regions, the efficiency of preconditioners in this case is demonstrated by some numerical experiments.

1 Introduction

Saddle point matrices arise in free Stokes flow and Darcy porous media flow problems as well as in the Brinkman flow, which couples Stokes and Darcy and appears in various environmental and technological applications. The considered saddle point matrices can be written in the form

$$\mathcal{A} = \begin{bmatrix} A & B^T \\ -B & 0 \end{bmatrix}, \tag{1}$$

where A is assumed to be symmetric and positive definite (SPD) and B has full (row) rank. A lot of effort was devoted to investigation of the preconditioners in cases of either Stokes or Darcy flow problems, in this paper we focus on unification of the results and considering Brinkman flow, which merges Stokes and Darcy cases. Brinkman flow has also a lot of applications ranging from power cells to e.g. flow and heat exchange in enhanced geothermal systems. The change of sign of the lower block matrix is convenient for definition of the investigated preconditioners. Note that with this change the Schur complement matrix $S = BA^{-1}B^T$ becomes SPD and the real part of the eigenvalues of $\mathcal{A}$ become positive.

© Springer International Publishing Switzerland 2016
T. Kozubek et al. (Eds.): HPCSE 2015, LNCS 9611, pp. 68–84, 2016.
DOI: 10.1007/978-3-319-40361-8_5

Since there is a strong variation of coefficients between different sub-domains such problems are ill-conditioned and the combination with the saddle point structure makes the construction of preconditioners more involved. In this paper we make a comparison of three classes of preconditioners, previously used for saddle point problems. Their analysis is simplified, extended and/or presented in a more uniform way than has appeared in the papers cited in sections devoted to particular preconditioners and the final secton.

For the iterative solution of systems with the matrices (1) we shall use both block diagonal and block triangular preconditioners. The standard form of the latter preconditioner takes the form

$$\mathcal{P} = \begin{bmatrix} A & 0 \\ -B & \tilde{S} \end{bmatrix} \tag{2}$$

for which a computation shows that

$$\begin{bmatrix} A & 0 \\ -B & \tilde{S} \end{bmatrix}^{-1} \begin{bmatrix} A & B^T \\ -B & 0 \end{bmatrix} = \begin{bmatrix} I & A^{-1}B^T \\ 0 & \tilde{S}^{-1}S \end{bmatrix},$$

where $S = BA^{-1}B^T$ is the Schur complement of A. Assuming no round-off errors, for such a matrix an optimal type of Krylov subspace iteration method, such as GMRES or GCG, converges to exact solution in $m+1$ iterations, where m is the number of distinct eigenvalues of $\tilde{S}^{-1}S$ not equal to unity. If $\tilde{S} = S$, it converges in two iterations.

Systems with A and $\tilde{S}$ (possibly then with $\tilde{S} = S$) can be solved by inner iterations to some accuracy, in which case some further outer iterations are needed to obtain a required tolerance.

As it is well-known, in some problems such as for the Stokes flow, systems with matrix A (i.e. the discretized Laplacian matrix) can be solved readily and the approximation $\tilde{S}$ can be chosen as a mass matrix, see e.g. [7]. However, for coupled free and porous media flow problems, matrix A becomes highly ill-conditioned due to discontinuous coefficients between Stokes and Darcy flow regions and heterogeneous coefficients appear in the Darcy region. The construction of an efficient preconditioner becomes then a more involved task.

In this paper, three types of preconditioning methods, which have been used earlier for simpler forms of saddle point problems, are compared. They include:

(i) The HSS method based on a Hermitian and skew-Hermitian splitting and regularization. This preconditioning, which uses just $\alpha I + BB^T$ instead of more complex Schur complement $S = BA^{-1}B^T$, is described and a simlified analysis is presented in Sect. 3.

(ii) Two block matrix preconditioners involving an approximation of the inverse of the block A and of the Schur complement matrix S are analyzed in Sect. 4. The analysis concerns triangular preconditioner as well as a special triangular preconditioner with $S_D = BDB^T$ approximation to the Schur complement S. The numerical experiments involve several types of D as an approximation to the inverse A^{-1}.

(iii) A block triangular factorization based on augmentation of the matrix A. Some variants of the preconditioner, again avoiding the work with the Schur complement matrix S, are introduced and analysed in Sect. 5.

In the next section we present shortly the mathematical models used. The preconditioners, their major properties and some numerical tests are presented in Sects. 3, 4 and 5. The final section summarizes the major properties of the preconditioners with some concluding remarks.

2   Coupled Free Flow and Porous Media Flow

A coupled fluid flow regime can be modelled by the equation

$$\left.\begin{array}{r} -\nu^\star \triangle u + \nu K^{-1} u + \nabla p = f \\ \nabla \cdot u = g \end{array}\right\} \text{ in } \Omega, \tag{3}$$

where u is the fluid velocity and p is the fluid pressure. The parameters involve the Brinkman parameter $\nu^\star$ (effective viscosity), fluid viscosity ν and permeability K. In our test problems we let $\nu^\star = \nu$. The parameters can vary in Ω in different ways, sometimes we can distinguish three regions $\Omega = \Omega_S \cup \Omega_B \cup \Omega_D$, where Ω_S represents the Stokes region with $K^{-1}u$ negligible regarding the other terms, Ω_D represents the Darcy region with $\nu^\star \triangle u$ negligible related to the Darcy term, and finally, Ω_B represents Brinkman region, where none of the terms in (3) is negligible. A model problem in a unit square with Stokes, Brinkman and Darcy regions can be seen in Fig. 1. The boundary conditions in the model problem are set to be

$$u = 0 \quad \text{on the bottom and top sides,}$$
$$p - \nu \frac{\partial u_n}{\partial n} = P_i \text{ and } \frac{\partial u_\tau}{\partial n} = 0 \quad \text{on the left and right sides.} \tag{4}$$

Above, $n \in \mathbb{R}^2$ denotes the unit outer normal vector, $u_n = u \cdot n$, $u_\tau = u \cdot \tau$, $\tau \in \mathbb{R}^2$ is a unit vector orthogonal to n. In our case such boundary conditions prescribe averaged pressure on the vertical sides, see e.g. [8].

Assuming $\nu^\star \geq \nu_0^\star > 0$, the equation can be variationally formulated in $\left[H^1(\Omega)\right]^d \times L_2(\Omega)$, where $\Omega \subset \mathbb{R}^d$ ($d = 2, 3$) and discretized by Taylor-Hood ($\mathbf{Q}_2 - \mathbf{Q}_1$) elements, see e.g. [7,18]. It leads to a finite element system with matrix (1) with $A = A_S + A_D$, where

$$\langle Au, v \rangle = \langle A_S u, v \rangle + \langle A_D u, v \rangle = \int_\Omega \nu^\star \nabla \bar{u} : \nabla \bar{v} \, dx + \int_\Omega \nu K^{-1} \bar{u} \cdot \bar{v} \, dx,$$

$$\langle Bu, p \rangle = \int_\Omega \text{div}(\bar{u}) \bar{p} \, dx,$$

where u, p are arbitrary vectors, $u \in \mathbb{R}^n$ and $p \in \mathbb{R}^m$ and $\bar{u}$ and $\bar{p}$ are the corresponding finite element functions. Note that in the sequel, we shall also

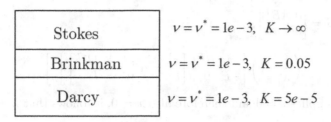

Stokes	$v = v^* = 1e-3, \ K \to \infty$
Brinkman	$v = v^* = 1e-3, \ K = 0.05$
Darcy	$v = v^* = 1e-3, \ K = 5e-5$

Fig. 1. A model problem in a unit square domain with three subregions and the parameter values used. Note that $K \to \infty$ means that the Darcy term is vanishing. The boundary conditions are $u = 0$ on the top and bottom sides, prescribed averaged pressure on the vertical sides.

deal with the pressure mass matrix M_p (the index p distinguish M and M_p), $\langle M_p p, q \rangle = \int_\Omega \overline{p} q \, dx$. Note that the formulation in $\left[H^1(\Omega) \right]^d \times L_2(\Omega)$ would be not correct for $v^* = 0$, i.e. for pure Darcy problem.

3 The HSS Preconditioning

In the HSS method one makes use of a splitting of $\mathcal{A}$ in a Hermitian and a skew–Hermitian term,

$$\mathcal{A} = \mathcal{H} + \mathcal{S}$$

where $\mathcal{H} = \begin{bmatrix} A & 0 \\ 0 & 0 \end{bmatrix}, \mathcal{S} = \begin{bmatrix} 0 & B^T \\ -B & 0 \end{bmatrix}.$

We assume here that A is positive definite and $0 < a_1 \leq \frac{x^T A x}{x^T x} \leq a_0,\ 0 < \sigma_1^2 \leq \frac{y^T B B^T y}{y^T y} \leq \sigma_0^2, \forall x \in \mathbb{R}^n, y \in \mathbb{R}^m, x \neq 0, y \neq 0$. For latter use, we assume that the matrices A and B are scaled so that a_0 and σ_0 are close to unit value. The preconditioner is defined as

$$\mathcal{P} = \frac{1}{\alpha}(\mathcal{H} + \alpha I)(\mathcal{S} + \alpha I) = \frac{1}{\alpha}\mathcal{H}\mathcal{S} + \mathcal{H} + \mathcal{S} + \alpha I, \qquad (5)$$

where $\alpha > 0$ is a parameter to be chosen.

To find the eigenvalues λ of the preconditioned matrix $\mathcal{P}^{-1}\mathcal{A}$, we consider the generalized eigenvalue problem,

$$\lambda \mathcal{P} \begin{bmatrix} x \\ y \end{bmatrix} = \mathcal{A} \begin{bmatrix} x \\ y \end{bmatrix}.$$

The derivation of the spectrum in [17] can be somewhat simplified. Since both $\mathcal{P}$ and $\mathcal{A}$ are nonsingular, λ exists and $\lambda \neq 0$. It holds

$$\lambda \left(\frac{1}{\alpha}\mathcal{H}\mathcal{S} + \mathcal{H} + \mathcal{S} + \alpha I \right) \begin{bmatrix} x \\ y \end{bmatrix} = (\mathcal{H} + \mathcal{S}) \begin{bmatrix} x \\ y \end{bmatrix}, \qquad \|x\| + \|y\| \neq 0$$

or, using $\mathcal{H}S = \begin{bmatrix} 0 & AB^T \\ 0 & 0 \end{bmatrix}$,

$$\left(\frac{1}{\lambda} - 1\right) \begin{bmatrix} A & B^T \\ -B & 0 \end{bmatrix} \begin{bmatrix} x \\ y \end{bmatrix} = \alpha \begin{bmatrix} I & \frac{1}{\alpha^2} AB^T \\ 0 & I \end{bmatrix} \begin{bmatrix} x \\ y \end{bmatrix}. \tag{6}$$

Since $\lambda = 1$ implies $y = 0$ and therefore also $x = 0$, it follows that $\lambda \neq 1$. By use of

$$\begin{bmatrix} I & -\frac{1}{\alpha^2} AB^T \\ 0 & I \end{bmatrix} \begin{bmatrix} A & B^T \\ -B & 0 \end{bmatrix} = \begin{bmatrix} A + \frac{1}{\alpha^2} AB^T B & B^T \\ -B & 0 \end{bmatrix},$$

(6) can be written in the form

$$\begin{bmatrix} AN & B^T \\ -B & 0 \end{bmatrix} \begin{bmatrix} x \\ y \end{bmatrix} = \eta \begin{bmatrix} x \\ y \end{bmatrix} \tag{7}$$

where $N = I + \frac{1}{\alpha^2} B^T B$ and $\eta = \frac{\alpha}{\frac{1}{\lambda}-1}$. Hence $\lambda = 1/(1 + \frac{\alpha}{\eta})$. Since $\lambda \neq 1$ and $\alpha > 0$, it follows that η exists and $\eta \neq 0$. Hence the second equation in (7) provides $y = -\frac{1}{\eta} Bx$, which after substitution into the first equation and multiplication of the arising equation by η gives

$$\eta ANx - B^T Bx = \eta ANx - \alpha^2 (N - I)x = \eta^2 x.$$

Let us take $z = \|Nx\|^{-1} Nx$, then the substitution to the previous equation for x and multiplication of the obtained equation by the Hermite transpose z^* provides

$$\eta(z^* Az) - \alpha^2 \left[z^* z - z^* N^{-1} z \right] = \eta^2 z^* N^{-1} z. \tag{8}$$

We use here the Hermitian product with the (complex) conjugate vector z^*, as $\begin{bmatrix} x \\ y \end{bmatrix}$ is the eigenvector of nonsymmetric matrix and therefore possibly complex. The Eq. (8) is a quadratic equation for η, which allows to express

$$\eta_{1,2} = \frac{1}{2c} (a \pm \sqrt{a^2 - 4\alpha^2 c(1 - c)}), \tag{9}$$

where $a = z^* Az$ and $c = z^* N^{-1} z$. This equation with the relation between η and λ shows that the eigenvalues of the preconditioned system $\mathcal{P}^{-1}\mathcal{A}$ are real if $a \geq 2\alpha\sqrt{c(1 - c)}$, i.e. $a \geq \alpha$ since $0 < c < 1$. There is no explicit formula for choosing an optimal value of α but it is seen that a small value of α gives a tighter spectrum around zero and one, see e.g. [1,17].

The implementation of $\mathcal{P}$, i.e. solution of linear systems with $\mathcal{P}$, requires solving a system with $A + \alpha I$ for the first factor in (5) and the Schur complement matrix $\alpha I + \frac{1}{\alpha} BB^T$ for the second factor $S + \alpha I$. In this case, we avoid the work with the more complex Schur complement $BA^{-1}B^T$. The second system $\alpha I + \frac{1}{\alpha} BB^T$ can be assembled, which permits use of an incomplete factorization method, although the assembled matrix is denser in the case of $\mathbf{Q}_2 - \mathbf{Q}_1$ elements. Therefore an iterative matrix-free (avoiding the assembling) solution of the second systems can be efficient. The efficiency of the preconditioner can be

Table 1. The numbers of iterations for solving the FEM system for the model problem from Fig. 1 by GMRES (relative residual accuracy for transformed residuals $\varepsilon = 1e{-}6$) with HSS preconditioner (5). The table shows dependence of the number of iterations on parameter α and the mesh size. Both SPD inner systems with the matrices $\alpha I + A$ and $\alpha I + \alpha^{-1} BB^T$ are solved by a direct solver.

Discretization $/\alpha$	1	1e−1	1e−3	1e−5
16×16	971	648	65	41
32×32	1772	862	78	59
64×64	2771	1901	112	91

seen from Table 1, which shows numbers of necessary iterations for solving the model problem from Fig. 1. From Table 1, it can be seen that relatively small α seems to be optimal. Unfortunately, even for small values of α, the preconditioner is not scalable with respect to the mesh size.

4 A Modified Block Triangular Matrix Preconditioner

Block type preconditioning for the saddle point matrix $\mathcal{A}$ can be constructed as an approximate version of the factorization

$$\mathcal{A} = \begin{bmatrix} A & B^T \\ -B & 0 \end{bmatrix} = \begin{bmatrix} A & 0 \\ -B & S \end{bmatrix} \begin{bmatrix} I & A^{-1}B^T \\ 0 & I \end{bmatrix},$$

where it is assumed that A is nonsingular, B has full rank and $S = BA^{-1}B^T$ is the Schur complement. The first factor used as a left preconditioner $\mathcal{P}$, provide the preconditioned system

$$\mathcal{P}^{-1}\mathcal{A} = \begin{bmatrix} A & 0 \\ -B & S \end{bmatrix}^{-1} \begin{bmatrix} A & 0 \\ -B & S \end{bmatrix} \begin{bmatrix} I & A^{-1}B^T \\ 0 & I \end{bmatrix} = \begin{bmatrix} I & A^{-1}B^T \\ 0 & I \end{bmatrix}$$

which has a minimal polynomial of order two. Introducing approximate Schur complement $\tilde{S} \sim S_A$ provides the preconditioner $\tilde{\mathcal{P}}$

$$\tilde{\mathcal{P}}^{-1}\mathcal{A} = \begin{bmatrix} A & 0 \\ -B & \tilde{S} \end{bmatrix}^{-1} \begin{bmatrix} A & B^T \\ -B & 0 \end{bmatrix} = \begin{bmatrix} A^{-1} & 0 \\ \tilde{S}^{-1}BA^{-1} & \tilde{S}^{-1} \end{bmatrix} \begin{bmatrix} A & B^T \\ -B & 0 \end{bmatrix} =$$

$$= \begin{bmatrix} I & A^{-1}B^T \\ 0 & \tilde{S}^{-1}BA^{-1}B^T \end{bmatrix} = \begin{bmatrix} I & A^{-1}B^T \\ 0 & \tilde{S}^{-1}S \end{bmatrix}. \tag{10}$$

In [1], we prove that the degree of the minimal polynomial for the matrix

$$\mathcal{X}_X = \begin{bmatrix} I & C \\ 0 & J \end{bmatrix}, \tag{11}$$

where $J \neq 0$ and $J - I$ is nonsingular, does not depend on C. Hence it equals the minimal polynomial degree for $\begin{bmatrix} I & 0 \\ 0 & J \end{bmatrix}$. The preconditioned matrix in (10) has the form of $\mathcal{X}_X$. Hence the degree of the minimum polynomial does not depend on the off-diagonal block matrix $A^{-1}B^T$, but only on the matrix $\begin{bmatrix} I & 0 \\ 0 & \tilde{S}^{-1}S \end{bmatrix}$. If $\tilde{S} = S$, i.e. $\tilde{S}^{-1}S = I$, then the degree of the minimal polynomial is 2 and it suffices to solve the corresponding system with two iterations of a Krylov subspace iteration method.

However, one must be aware of that round–off errors may cause perturbations that can increase the number of iterations significantly over that of the minimal degree. For a discussion of the influence of less accurate computations, see [9].

The above indicates that even if we do not solve the arising systems with matrix A in the preconditioners exactly, but accurately enough, there is no need to include the second factor in the preconditioner to decrease the number of iterations further. However, as said, in general when these systems are solved less accurately it may cause some increase of the number of iterations. The use of the following modified block triangular factorization can dampen this increase.

A major problem with factorization of saddle point matrices is the lack of accurate approximations of the Schur complement matrix, that are readily implemented. For the construction of a Schur complement matrix approximation, let us assume that D is an approximate inverse of the pivot block matrix, $D \approx A^{-1}$. Then D can be used for definition of the preconditioner $\mathcal{P}_D$ to $\mathcal{A}$,

$$\mathcal{P} = \mathcal{P}_D = \begin{bmatrix} A & ADB^T \\ -B & 0 \end{bmatrix} = \begin{bmatrix} A & 0 \\ -B & S_D \end{bmatrix} \begin{bmatrix} I & DB^T \\ 0 & I \end{bmatrix} \tag{12}$$

where $S_D = BDB^T$. A computation shows that

$$\mathcal{P}_D^{-1}\mathcal{A} = \begin{bmatrix} I & -DB^T \\ 0 & I \end{bmatrix} \begin{bmatrix} I & 0 \\ 0 & S_D^{-1}S_A \end{bmatrix} \begin{bmatrix} I & A^{-1}B^T \\ 0 & I \end{bmatrix} =$$

$$= \begin{bmatrix} I & -DB^T \\ 0 & I \end{bmatrix} \begin{bmatrix} I & A^{-1}B^T \\ 0 & S_D^{-1}S_A \end{bmatrix} = \begin{bmatrix} I & A^{-1}B^T - DB^T S_D^{-1}S_A \\ 0 & S_D^{-1}S_A \end{bmatrix} =$$

$$= \begin{bmatrix} I & A^{-1}E_{11}B^T - B_{12}E_{22} \\ 0 & I + E_{22} \end{bmatrix},$$

where $E_{11} = I - AD$, $B_{12} = DB^T$, $E_{22} = S_D^{-1}S_A - I$, $S_A = BA^{-1}B^T$. Here

$$E_{22} = S_D^{-1}B(D + (I - DA)A^{-1})B^T - I = S_D^{-1}B(I - DA)A^{-1}B^T,$$

so with an accurate approximation D of A^{-1}, both $\|E_{11}\|$ and $\|E_{22}\|$ are small.

The matrix has the same form (11) as is discussed earlier. Therefore, for an accurate solution where the minimal polynomial for $I + E_{22}$ is nearly reached, the off-diagonal matrix has little influence on the number of iterations. Even when one interrupts the iterations at an earlier stage, this matrix has little influence when $\|E_{11}\|$ and $\|E_{22}\|$ are small. This holds at least if one uses multiple precision computation.

Note that the spectrum of $\mathcal{P}_D^{-1}\mathcal{A}$ consists of unit eigenvalues corresponding to eigenvectors $[x, 0]$, and of eigenvalues of $S_D^{-1}S_A$, i.e.

$$\sigma\left(\mathcal{P}_D^{-1}\mathcal{A}\right) = \{1, \sigma(S_D^{-1}S_A)\}.$$

Thus the spectrum appears favourable for the convergence of iterative methods, but $\mathcal{P}_D^{-1}\mathcal{A}$ is not normal.

At each iteration, the preconditioning involves solving systems with A (possibly solved by some inner iteration with the preconditioner D, i.e. approximation of its inverse) and solution of systems with S_D. This latter can take place by forming S_D explicitly and solving the corresponding system with a suitable direct or iterative solution method.

For the Stokes problem with $A = A_S$, the Schur complement $S = BA^{-1}B^T$ can be efficiently preconditioned by the pressure mass matrix M_p. For the Darcy problem with $A = A_D$, the approximate Schur complement S_D can use $D = \text{diag}(A)^{-1}$. In the case of Brinkman problem $\Omega = \Omega_B$ with constant coefficients a combination of the above approximations can be used as in the generalized Stokes problem, see [6,13].

For the investigated case, when $\Omega = \Omega_S \cup \Omega_B \cup \Omega_D$ some other techniques using a sparse approximate inverse $D \sim A^{-1}$ were tested, see the following list

(a) $D = D_{diag} = \text{diag}(A)^{-1}$ is the simplest choice,

(b) $D = D_{el} = \sum_e R_e^T A_e^{-1} R_e$ where $e \in T_h$, R_e is the restriction to degrees of freedom of e and A_e is a regularized local finite element matrix (regularization by $h^2 I_e$ is used for $e \in \Omega_S$, cf. [14]),

(c) $D = D_{as} = \sum_e R_e^T \bar{A}_e^{-1} R_e$ where $\bar{A}_e = R_e A R_e^T$ is a submatrix of the fully assembled finite element matrix A, see [15]. The matrix $\bar{A}_e$ can be also viewed as the finite element matrix assembled for macroelement $E = E_e \subset \Omega \subset \mathbb{R}^d$ with zero Dirichlet condition on $\partial E \backslash (\partial \Omega \backslash \partial_D \Omega)$. Here the macroelement $E = E_e$ is the union of all elements $t \in T_h$ that share with $e \in T_h$ some degree(s) of freedom, $\partial_D \Omega$ is the part of the domain Ω where Dirichlet boundary condition is prescribed. From the latter point of view D_{as} is an additive Schwarz type approximation to A^{-1},

(d) both $D = D_{as}$ and $D = D_{el}$ can be easily improved by a diagonal correction X such that the Frobenius norm $\|I - (D + X)A\|_F = \|R - XA\|_F$ is minimized, see [15]. Note that it can be easily shown that the minimization is given by X such that $X_{ii} = (AR^T)_{ii}/(AA^T)_{ii}$. The corrected D is denoted as D_{el}^{F1} and D_{as}^{F1}, respectively.

(e) an even cheaper diagonal correction X can be obtained by minimizing a weighted Frobenius norm $\|I - (D + X)A\|_W$, $W = A^{-1}$. Then the minimization provides X such that $X_{ii} = (\sum_j A_{ij})/A_{ii}$. This technique is efficient for correcting D_{as}, see D_{as}^{F2} in Table 1.

(f) $D = D_{ainv}$ - the weighted Frobenius norm $\|\cdot\|_W$, where W is a nonsingular matrix, can be used for construction of various sparse approximate inverses to A^{-1}, see [2], Chap. 8. Here we shall use D_{ainv} which is SPD and has the same sparsity pattern as A. The $D = D_{ainv} = LL^T$ is computed by the technique of Kolotilina and Yeremin, see e.g. [2,11].

Note that D_{as} could be improved by taking larger macroelements E and/or by adding a coarse space term to the expression for D_{as} (two-level additive Schwarz), but this deteriorates the sparsity of D_{as} and can be considered only

Table 2. The numbers of iterations for solving the system with matrix A for the model problem from Fig. 1. The iterations use PCG with relative residual accuracy $\varepsilon = 1e{-}6$.

Discretization/$D =$	D_{diag}	D_{as}	D_{ainv}
16×16	50	46	30
32×32	99	91	60
64×64	188	171	119

Table 3. The numbers of iterations for solving the FEM system for the model problem from Fig. 1 by GMRES (relative residual accuracy for transformed residuals $\varepsilon = 1e{-}6$) with the block preconditioner (12). The table shows dependence on various choices of matrix D and the mesh size. Both SPD inner systems with matrices A and S_D are solved by a direct solver.

Discretization/$D =$	D_{diag}	D_{el}	D_{el}^{F1}	D_{as}	D_{as}^{F1}	D_{as}^{F2}	D_{ainv}
16×16	31	39	24	20	40	19	18
32×32	48	38	29	30	33	28	27
64×64	64	39	39	46	39	42	40

if the system with S_D is solved iteratively. In this case, we could also use an algebraic multigrid preconditioning method as an alternative to the additive Schwarz technique.

The simple approximate inverses described above can serve as preconditioner to A (see Table 2) and as a tool for constructing approximate Schur comlement, which can be used for the solution of Brinkman problem (see Table 3).

Tables 2 and 3 show the efficiency of selected approximate inverse preconditioners. With mesh refinement, the number of iterations grows as rapidly as with the unpreconditioned Krylov type iterative method (factor 2 in Table 2). The number of outer iterations in Table 3 grows less rapidly (about factor $\sqrt{2}$). In comparison with HSS preconditioning (Table 1) the number of inner iterations are fewer, but the inner systems are somewhat more complicated. Instead of system with the matrix $\alpha I + \frac{1}{\alpha} BB^T$, we solve the system with $S_D = BDB^T$, which is more difficult to compute and which moreover requires computation of $D \sim A^{-1}$. On the other hand, the computed D can be exploited as preconditioner for solving the second inner system with the matrix A. Table 3 indicates also that use of exact local approximations D_{el} or additive Schwarz type preconditioner D_{as}, give a mesh independent number of iterations. As shown in [4], such preconditioners can indeed be more efficient than simply use of the corresponding coarse mesh matrix. They are somewhat less sparse but can be efficiently factorized in linear time.

5   Augmented Matrix Type Preconditioners

There are two versions of this method. Both are based on the augmented pivot block matrix, $A_W = A + B^T W^{-1} B$, in particular on $A_r = A + r B^T W^{-1} B$ when W is replaced by $\frac{1}{r} W$ and on various choices of W.

5.1 A Full Block Factorization Preconditioner Based on an Augmented Matrix

For reasons to be explained at the end of this subsection, we consider the augmented matrix preconditioner for the generalized saddle point matrix, $\mathcal{A} = \begin{bmatrix} A & B^T \\ -B & C \end{bmatrix}$, where A and C are symmetric and positive semidefinite (SPSD), but at least one of them is SPD and B has full rank. Further assume that $\mathcal{N}(A) \cap \mathcal{N}(B) = \{0\}$.

When A is ill–conditioned and/or it is difficult to approximate the Schur complement accurately, one must use a preconditioner based on an augmented matrix; $A_W = A + B^T W^{-1} B$, where W is SPD. In practice W or W^{-1} is a sparse matrix. To enable solving arising systems with A_W one can try to find proper choices of W such as an approximate inverse of BB^T or $BA^{-1}B^T$. As we shall see, use of such an augmented matrix avoids the construction of the Schur complement matrix $S = BA^{-1}B^T$, which is difficult when A is ill-conditioned. However, the difficulty lies now in the solution of systems with the augmented matrix A_W. Mostly one applies some inner iteration method for solving systems with A_W.

Let $\mathcal{P} = \begin{bmatrix} A & B^T \\ -B & W \end{bmatrix}$ be a preconditioner to $\mathcal{A}$. It holds

$$\mathcal{P} = \begin{bmatrix} A_W & B^T \\ 0 & W \end{bmatrix} \begin{bmatrix} I & 0 \\ -W^{-1}B & I \end{bmatrix}$$

and

$$\mathcal{P}^{-1} = \begin{bmatrix} I & 0 \\ W^{-1}B & I \end{bmatrix} \begin{bmatrix} A_W^{-1} & -A_W^{-1}B^T W^{-1} \\ 0 & W^{-1} \end{bmatrix}.$$

Hence a solution of a system with $\mathcal{P}$ requires just one solution with A_W and two solutions of a system with W or matrix vector multiplications with W^{-1}.

Further it holds

$$\mathcal{P}^{-1}\mathcal{A} = I + \mathcal{P}^{-1}(\mathcal{A} - \mathcal{P}) = I + \mathcal{E},$$

where

$$\mathcal{E} = \begin{bmatrix} E_{11} & E_{12} \\ E_{21} & E_{22} \end{bmatrix} = \begin{bmatrix} I & 0 \\ W^{-1}B & I \end{bmatrix} \begin{bmatrix} A_W^{-1} & -A_W^{-1}B^T W \\ 0 & W^{-1} \end{bmatrix} \begin{bmatrix} 0 & 0 \\ 0 & C - W \end{bmatrix} =$$

$$= \begin{bmatrix} I & 0 \\ W^{-1}B & I \end{bmatrix} \begin{bmatrix} 0 & -A_W^{-1}B^T(W^{-1}C - I) \\ 0 & W^{-1}C - I \end{bmatrix} =$$

$$= \begin{bmatrix} 0 & -A_W^{-1}B^T(W^{-1}C - I) \\ 0 & (I - W^{-1}BA_W^{-1}B^T)(W^{-1}C - I) \end{bmatrix}.$$

Here, by use of the Sherman-Morrison-Woodbury identity,

$$(BA_W^{-1}B^T)^{-1} = (B(A + B^T W^{-1}B)^{-1}B^T)^{-1} = W^{-1} + (BA^{-1}B^T)^{-1}. \quad (13)$$

Hence

$$
\begin{aligned}
I - W^{-1}BA_W^{-1}B^T &= I - W^{-1}(W^{-1} + (BA^{-1}B^T)^{-1})^{-1} = \\
&= I - [I + (W^{-1}BA^{-1}B^T)^{-1}]^{-1} = \\
&= I - W^{-1}BA^{-1}B^T(I + W^{-1}BA^{-1}B^T)^{-1} = \\
&= (I + W^{-1}BA^{-1}B^T)^{-1}.
\end{aligned}
\tag{14}
$$

We can replace W with $W_r = rW$, where $r \geq 1$ is a parameter. This makes the matrix in (14) equal to $(I + rW^{-1}BA^{-1}B^T)^{-1}$ where $W^{-1}BA^{-1}B^T$ is non-singular, and hence the matrix block E_{22} is $O(1/r)$, i.e. small for large values of r and all eigenvalues cluster at unity. Such approximation is accurate for large values of r, a Krylov subspace method will require very few, typically 2 to 4 iterations, assuming sufficiently small round–off errors.

It is seen that $W = BA^{-1}B^T$ is a good choice. Since this matrix is unavailable in practice, we can instead let W approximate BB^T. This makes $A_W = A + B^TW^{-1}B$ equal to A plus an approximation of a projection matrix. If C is SPD, we can alternatively let W be an approximation of C. If W is an accurate approximations of C, then both E_{12} and E_{22} are small and few iterations are needed here also.

5.2   A Special Block Triangular Preconditioner

For the original matrix $\mathcal{A} = \begin{bmatrix} A & B^T \\ B & 0 \end{bmatrix}$ we can instead use the block triangular preconditioner, $\begin{bmatrix} A_r & 2B^T \\ 0 & -W_r \end{bmatrix}$, where $A_r = A + rB^TW^{-1}B$ is an augmented matrix. Although this matrix seems to have little in common with $\mathcal{A}$ we shall see that it leads a suitable spectrum for fast rate of converge.

The corresponding generalized eigenvalue problem,

$$
\lambda \begin{bmatrix} A_r & 2B^T \\ 0 & -W_r \end{bmatrix} \begin{bmatrix} x \\ y \end{bmatrix} = \begin{bmatrix} A & B^T \\ B & 0 \end{bmatrix} \begin{bmatrix} x \\ y \end{bmatrix}
$$

where $W_r = \frac{1}{r}W$, $r \geq 1$ and A is assumed to be SPD, is transformed by a similarity transformation, which for the generalized eigenvalue problem corresponds to a multiplication of the matrices from both sides with $\begin{bmatrix} A_r^{-1/2} & 0 \\ 0 & W_r^{-1/2} \end{bmatrix}$.

Since $A = A_r - B^TW_r^{-1}B$, this leads to

$$
\lambda \begin{bmatrix} I & 2F^T \\ 0 & -I \end{bmatrix} \begin{bmatrix} \tilde{x} \\ \tilde{y} \end{bmatrix} = \begin{bmatrix} I - F^TF & F^T \\ F & 0 \end{bmatrix} \begin{bmatrix} \tilde{x} \\ \tilde{y} \end{bmatrix}
$$

where $F = W_r^{-1/2}BA_r^{-1/2}$; and $\tilde{x} = A_r^{1/2}x$, and $\tilde{y} = W_r^{1/2}y$.

Since,

$$
\begin{bmatrix} I & 2F^T \\ 0 & -I \end{bmatrix}^{-1} = \begin{bmatrix} I & 2F^T \\ 0 & -I \end{bmatrix},
$$

it holds

$$\lambda \begin{bmatrix} \widetilde{x} \\ \widetilde{y} \end{bmatrix} = \begin{bmatrix} I & 2F^T \\ 0 & -I \end{bmatrix} \begin{bmatrix} I - F^T F & F^T \\ F & 0 \end{bmatrix} \begin{bmatrix} \widetilde{x} \\ \widetilde{y} \end{bmatrix} \quad \text{or} \quad \lambda \begin{bmatrix} \widetilde{x} \\ \widetilde{y} \end{bmatrix} = \begin{bmatrix} I + F^T F & F^T \\ -F & 0 \end{bmatrix} \begin{bmatrix} \widetilde{x} \\ \widetilde{y} \end{bmatrix}.$$

Here

$$\begin{bmatrix} I + F^T F & F^T \\ -F & 0 \end{bmatrix} \begin{bmatrix} I & -F^T \\ -F & I \end{bmatrix} = \begin{bmatrix} I & -F^T \\ -F & I \end{bmatrix} \begin{bmatrix} I & 0 \\ 0 & FF^T \end{bmatrix},$$

i.e.,

$$\begin{bmatrix} I & -F^T \\ -F & I \end{bmatrix} \quad \text{and} \quad \begin{bmatrix} I & 0 \\ 0 & FF^T \end{bmatrix}$$

are the eigenvector and eigenvalue matrices, respectively. It follows that the multiplicity of the eigenvalue $\lambda = 1$ is at least n, if A has order $n \times n$. It is easily seen that the preconditioned matrix is diagonalizable.

Further $FF^T = W_r^{-1/2} B A_r^{-1} B^T W_r^{-1/2}$ so, by use of (13), follows the useful relation,

$$(FF^T)^{-1} = W_r^{1/2} (B A_r^{-1} B^T)^{-1} W_r^{1/2} =$$
$$= W_r^{1/2} (W_r^{-1} + (B A^{-1} B^T)^{-1}) W_r^{1/2} =$$
$$= I + W_r^{1/2} (B A^{-1} B^T)^{-1} W_r^{1/2}.$$

Therefore $\sigma(FF)^T = \frac{1}{1+1/(r\mu)}$, where μ is an eigenvalue of $\mu W z = B A^{-1} B^T z$, $z \neq 0$. Since B has full rank, it follows that $\mu \neq 0$, so $\mu > 0$.

Hence, W should indeed be related to the Schur complement matrix $B A^{-1} B^T$. However, for large values of r the actual choice has less influence on the spectrum of the preconditioned matrix.

As we have shown, the preconditioned matrix is diagonalizable. As $r \to \infty$, all eigenvalues cluster at $\lambda = 1$. The eigenvalue bounds, $\frac{1}{(1+\frac{1}{r\mu_{min}})} \leq \lambda \leq 1$ can be controlled by choosing the parameter r sufficiently large. Unless μ_{min} is extremely small this leads to an efficient iteration method, such as a preconditioned form of GMRES or GCG. However, in practice, the arising systems with A_r are solved by an inner iteration method, perhaps with a variable stopping criteria, so a variable preconditioned version of the above methods (see [2, 16]) must be used.

Note that the second matrix term in $A_r = A + B^T W_r^{-1} B$ is singular so the spectrum of A_r may still contain part of the spectrum to A, namely if the corresponding eigenvectors of A belong to $\mathcal{N}(B)$.

We consider now methods to precondition the matrix A_r that appears in each outer iteration. If one chooses W as a diagonal matrix one can form the matrix $A_r = A + B^T W_r^{-1} B$ by elementwise assembly explicitly and use e.g. an ILU or AMG method to solve the systems with A_r. Even if W is not diagonal, one can form local elementwise versions of $B^T W^{-1} B$ exactly, to then be assembled, see e.g. [4, 12].

However, such a matrix W will in general be somewhat less efficient in getting a good spectrum of the matrix

$$FF^T = W_r^{-1/2} B A_r^{-1} B^T W_r^{-1/2} = I + \frac{1}{r} W^{-1/2} B A^{-1} B^T W^{-1/2} \tag{15}$$

since the second term may be ill–conditioned, i.e., it may give many outer iterations. It can be improved by choosing a larger value of the parameter r, but this
results in a more ill-conditioned matrix A_r.

A more efficient choice is $W = BB^T$ or $BD^{-1}B^T$, where D is some diagonal
matrix approximation of A. The choice $W = BB^T$ makes $BW^{-1}B^T$ a projection
matrix. For $W = BD^{-1}B^T$, the matrix $P = D^{-1}B^TW^{-1}B$ becomes a projection
matrix with $BP = B$, since $P^2 = P$.

If we choose $W = BA^{-1}B^T$ which, however, is infeasible, then the norm of
the second matrix in (15) would equal $1/r$, i.e. FF^T is very close to the identity
matrix for large values of r, resulting in few outer iterations. The choice $D =$
diagonal approximation of A is a compromise.

To solve systems with A_r we first multiply by D^{-1}, to get

$$\widetilde{A}_r = D^{-1}A_r = D^{-1}A + rD^{-1}B^TW^{-1}B.$$

This matrix is preconditioned with $(I+rP)^{-1}$, e.g. from right. Since $(I+rP)^{-1} =$
$I - r/(r + 1)P$, when P is a projection operator, it holds

$$\begin{aligned}
\widetilde{A}_r(I + rP)^{-1} &= (D^{-1}A + rD^{-1}B^TW^{-1}B)(I + rP)^{-1} = \\
&= (D^{-1}A - I + I + rP)(I + rP)^{-1} = \\
&= I + (D^{-1}A - I)(I - r/(r + 1)P).
\end{aligned} \tag{16}$$

Hence, the preconditioning is particularly efficient if $D^{-1}A - I$ has small actions
on vectors $x \in \mathcal{N}(B)$ where $Px = 0$. For vectors x in the orthogonal complement,
the projection factor results in $(I - r/(r + 1)P)x = 1/(r + 1)x$, i.e. a small
perturbation for large values of r. Further, if A is dominated by a mass matrix
as in the Darcy region and we let $D = [d_i]$, $d_i = (Ae)_i$, $e = (1, 1, \ldots, 1)^T$, then
$D^{-1}(A-D)$ becomes approximately a difference operator and for smooth vectors
x, $D^{-1}(A - D)x = O(h^2)$, where h is the mesh size in the difference or finite
element mesh. Hence $\widehat{A}x = x + O(h^2)$ for such vectors and this indicates a fast
convergence of the iteration method for solving systems with A_r.

5.3   A Method Based on Projections

Let $D = L^T L$ is a sparse, SPD approximation of A^{-1} such as e.g. the diagonal
D_{diag} and D_{ainv} or any other approximation introduced in the previous section.
Then for $W = \frac{1}{r}BDB^T$ the augmented matrix

$$A_r = A + rB^T(BDB^T)^{-1}B$$

can be approximated by

$$\begin{aligned}
\widetilde{A}_r &= D^{-1} + rB^T(BDB^T)^{-1}B = L^{-1}L^{-T} + rB^T(BDB^T)^{-1} \\
&= L^{-1}\left(I + rLB^T(BDB^T)^{-1}BL^T\right)L^{-T} \\
&= L^{-1}\left(I + rP\right)L^{-T},
\end{aligned}$$

where $P = LB^T(BDB^T)^{-1}BL^T$ is a projection matrix. Using the projection property, $P^2 = P$, systems with $(I + rP)$ can be solved directly in the following way. A multiplication of the equation $(I + rP)x = y$ by P and $I - P$ provides $(1 + r)Px = Py$, i.e. $Px = \frac{1}{1+r}Py$, and $(I - P)x = y - Py$. Thus

$$x = Px + (I - P)x = y - (1 - \frac{1}{1+r})Py.$$

It provides

$$(I + rP)^{-1} = I - \frac{r}{1+r}P$$

As a preconditioner to the matrix A_r, we now use $\tilde{A}_r$, where

$$\tilde{A}_r^{-1} = L^T (I + rP)^{-1} L = L^T \left(I - \frac{r}{1+r}P\right) L = D - \frac{r}{1+r}DB^T(BDB^T)^{-1}BD. \tag{17}$$

Here a computation of Py requires a matrix vector multiplication and a solution with a system with matrix $\tilde{W} = BDB^T$. The preconditioned matrix can be written in the form

$$L^T \left(I - \frac{r}{1+r}P\right) L \left(A + rB^T(BA^{-1}B^T)^{-1}B\right)$$

with the spectrum equal to the spectrum of the matrix

$$\left(I - \frac{r}{1+r}P\right) L \left(A + rB^T(BA^{-1}B^T)^{-1}B\right) L^T$$

$$= \left(I - \frac{r}{1+r}P\right) (LAL^T + rP) = I + \left(I - \frac{r}{r+1}P\right)(LAL^T - I).$$

It is seen that an iteration method with this matrix works efficiently if the residuals $(LAL^T - I)x$ have small components in $\mathcal{N}(B)$. The second factor damps the residual components in $\mathcal{N}(B)^\perp$ with a factor $1/(r + 1)$.

The performed numerical tests first use the simplest choice $D = \text{diag}(A)$. The outer iterations for solving the system with matrix $\mathcal{A}$ are performed by GMRES with the preconditioner $\mathcal{P}$,

$$\mathcal{A} = \begin{bmatrix} A & B^T \\ -B & 0 \end{bmatrix}, \quad \mathcal{P} = \begin{bmatrix} A_r & \zeta B^T \\ 0 & W_r \end{bmatrix}, \quad W_r = \frac{1}{r}BDB^T. \tag{18}$$

The system $A_r = A + rB^T(BDB^T)^{-1}B$ is solved by inner GMRES iterations with inner preconditioner $\tilde{A}_r$ introduced above. The application of $\tilde{A}_r^{-1}$ requires to solve one system with $W = BDB^T$. If this solution is done by direct solver, we get numerical results which are shown in Table 4.

Table 4. Table of numbers of outer GMRES iterations and in parentheses average number of inner GMRES iterations per one outer iteration. Outer iterations solve the system with matrix $\mathcal{A}$ and use preconditioner $\mathcal{P}$, see (18). Inner iterations solve the system with the augmented matrix $A_r = A + rB^T W^{-1} B$, where $W = BDB'$. The preconditioner $\tilde{A}_r$ is implemented via the projection technique (17). Application of $\tilde{A}_r^{-1}$ involves solution of the system with W, which is done by a direct solver. Application of $\mathcal{P}$ involves further solution of system with W, which is handled in the same way. Both inner and outer iterations were solved by GMRES method with $\varepsilon = 1e-6$.

	Mesh	r = 1e0			r = 1e3			r = 1e6		
		$\zeta = 0$	$\zeta = 1$	$\zeta = 2$	$\zeta = 0$	$\zeta = 1$	$\zeta = 2$	$\zeta = 0$	$\zeta = 1$	$\zeta = 2$
D_{diag}	16 × 16	18(46)	13(47)	8(46)	3(44)	5(44)	3(41)	2(44)	3(43)	2(45)
	32 × 32	17(77)	13(80)	7(85)	3(79)	5(77)	3(72)	2(79)	3(76)	2(81)
	64 × 64	15(121)	13(128)	7(145)	3(138)	5(137)	3(126)	2(137)	3(139)	2(144)
D_{as}	16 × 16	22(44)	13(45)	9(45)	3(41)	5(42)	3(41)	2(41)	3(41)	2(25)
	32 × 32	20(77)	13(80)	9(79)	3(79)	5(78)	3(76)	2(77)	3(77)	2(79)
	64 × 64	19(132)	15(132)	8(141)	3(149)	5(138)	3(139)	2(148)	3(146)	2(147)
D_{ainv}	16 × 16	18(25)	11(26)	8(26)	3(24)	5(25)	3(23)	2(24)	3(25)	2(42)
	32 × 32	18(45)	11(46)	8(48)	3(48)	5(47)	3(44)	2(47)	3(48)	2(47)
	64 × 64	16(80)	14(80)	11(88)	3(89)	5(86)	3(82)	2(90)	3(84)	2(92)

6   Summary of Properties and Concluding Remarks

The paper deals with three type of preconditioners, each of the presented methods has some advantages and disadvantages. First, we investigate the HSS method introduced by Bai et al. [5], see also an overview in [6] and later development in [17]. We present a simplified proof of the preconditioning effects and some numerical test which unfortunately does not scale with the mesh size. The advantage of this method are simplest inner block systems, disadvantage is in larger number of outer iterations. Secondly, we consider preconditioners based on the Schur complement $S = BA^{-1}B^T$ with a suitable approximation of the inverse of A. These methods are investigated for a longer time, see [6], but we present some new views on a standard triangular preconditioner, and work with a modified triangular preconditioner (12) which can be also viewed as incomplete factorization with two triangular factors. The numerical experiments show even numerical scalability for some choices of approximate inverse of A, which needs a further investigation. Thirdly, we analyse preconditioners based on augmentation of the pivot block A, see [10] as well as e.g. [3]. We outline analysis of the standard triangular preconditioner, triangular preconditioner with doubled off diagonal block and a version with the augmented system close to projection. As spectral information is not enough to control the convergence for nonsymmetric preconditioned systems, we also added some information about diagonalizability of the preconditioned system.

In Table 5 the major properties of the methods are listed. We refer here, in particular, to the coupled Stokes-Brinkman-Darcy problems.

Table 5. Summary of properties of the preconditioners

	Preconditioning method	Systems to be solved at each outer iteration	Condition number; inner iteration	Condition number; outer iteration
I	HSS	$\alpha I + A,$ $\alpha I + \frac{1}{\alpha} BB^T$	(1)	$\gg 1,$ clustered spectra near 0, and 1
II	Factorization of approximate saddle point	$A, (D)^{4)}, S_D$	(1)	$\kappa(S_D^{-1} S_A) \gg 1$
III	Augmented Matrix approach	$A + rB^T W^{-1} B, W$	(2)	$O(1)^3$

(1) Normally not extremely large, except when A is ill–conditioned, for instance due to heterogeneous material coefficients.

(2) Condition number of $A + rB^T W^{-1} B$ is large for large values of r but spectrum is clustered, typically in two intervals.

(3) Normally 2 to 4 iterations, depending on inner iteration accuracy. A relaxed stopping criteria for the inner iterations with A_r increases the number of outer iterations only mildly.

(4) If we use an explicit approximation D of A^{-1}, there will only occur matrix vector multiplications with D.

Acknowledgement. This work was supported by the European Regional Development Fund in the IT4Innovations Centre of Excellence project (identification number CZ.1.05/1.1.00/02.0070). The comments by referees are greatly acknowledged.

References

1. Axelsson, O., Blaheta, R.: Preconditioning methods for saddle point problems arising in porous media flow (submitted)
2. Axelsson, O.: Iterative Solution Methods. Cambridge University Press, Cambridge (1996)
3. Axelsson, O., Blaheta, R.: Preconditioning of matrices partitioned in 2 × 2 block form: Eigenvalue estimates and Schwarz DD for mixed FEM. Numer. Linear Algebra Appl. **17**(5), 787–810 (2010). http://dx.org/10.1002/nla.728
4. Axelsson, O., Blaheta, R., Neytcheva, M.: Preconditioning of boundary value problems using elementwise Schur complements. SIAM J. Matrix Anal. Appl. **31**(2), 767–789 (2009)
5. Bai, Z.Z., Golub, G.H., Ng, M.K.: Hermitian and skew-Hermitian splitting methods for non-Hermitian positive definite linear systems. SIAM J. Matrix Anal. Appl. **24**(3), 603–626 (2003)
6. Benzi, M., Golub, G.H., Liesen, J.: Numerical solution of saddle point problems. Acta Numerica **14**(1), 1–137 (2005)

7. Elman, H., Silvester, D., Wathen, A.: Finite Elements and Fast Iterative Solvers: With Applications in Incompressible Fluid Dynamics. Oxford University Press, Oxford (2014)
8. Formaggia, L., Gerbeau, J.F., Nobile, F., Quarteroni, A.: Numerical treatment of defective boundary conditions for the Navier-Stokes equations. SIAM J. Numer. Anal. **40**(1), 376–401 (2002)
9. Greenbaum, A., Strakos, Z.: Predicting the behavior of finite precision Lanczos and conjugate gradient computations. SIAM J. Matrix Anal. Appl. **13**(1), 121–137 (1992)
10. Greif, C., Schötzau, D.: Preconditioners for saddle point linear systems with highly singular (1,1) blocks. Electron. Tran. Numer. Anal. ETNA **22**, 114–121 (2006)
11. Kolotilina, L.Y., Yeremin, A.Y.: Factorized sparse approximate inverse preconditionings I. Theory. SIAM J. Matrix Anal. Appl. **14**(1), 45–58 (1993)
12. Kraus, J.: Algebraic multilevel preconditioning of finite element matrices using local Schur complements. Numer. Linear Algebra Appl. **13**(1), 49–70 (2006)
13. Mardal, K.A., Winther, R.: Uniform preconditioners for the time dependent Stokes problem. Numer. Math. **98**(2), 305–327 (2004)
14. Neytcheva, M.: On element-by-element Schur complement approximations. Linear Algebra Appl. **434**(11), 2308–2324 (2011)
15. Neytcheva, M., Bängtsson, E., Linnér, E.: Finite-element based sparse approximate inverses for block-factorized preconditioners. Adv. Comput. Math. **35**(2), 323–355 (2011)
16. Saad, Y.: Iterative Methods for Sparse Linear Systems. SIAM, Philadelphia (2003)
17. Simoncini, V., Benzi, M.: Spectral properties of the Hermitian and skew-Hermitian splitting preconditioner for saddle point problems. SIAM J. Matrix Anal. Appl. **26**(2), 377–389 (2004)
18. Taylor, C., Hood, P.: A numerical solution of the Navier-Stokes equations using the finite element technique. Comput. Fluids **1**(1), 73–100 (1973)

Efficient Implementation of Total FETI Solver for Graphic Processing Units Using Schur Complement

Lubomír Říha[1]([✉]), Tomáš Brzobohatý[1], Alexandros Markopoulos[1],
Tomáš Kozubek[1], Ondřej Meca[1], Olaf Schenk[2], and Wim Vanroose[3]

[1] IT4Innovations National Supercomputing Centre,
17. Listopadu 15/2172, Ostrava, Czech Republic
{lubomir.riha,tomas.brzobohaty,alexandros.markopoulos,
tomas.kozubek,ondrej.meca}@vsb.cz
[2] Institute of Computational Science, Universita della Svizzera italina,
Via Giuseppe Buffi 13, 6900 Lugano, Switzerland
olaf.schenk@usi.ch
[3] University of Antwerp, Department of Mathematics and Computer Science,
Middelheimlaan 1, 2020 Antwerp, Belgium
wim.vanroose@uantwerpen.be

Abstract. This paper presents a new approach developed for acceleration of FETI solvers by Graphic Processing Units (GPU) using the Schur complement (SC) technique. By using the SCs FETI solvers can avoid working with sparse Cholesky decomposition of the stiffness matrices. Instead a dense structure in form of SC is computed and used by conjugate gradient (CG) solver. In every iteration of CG solver a forward and backward substitution which are sequential are replaced by highly parallel General Matrix Vector Multiplication (GEMV) routine. This results in 4.1 times speedup when the Tesla K20X GPU accelerator is used and its performance is compared to a single 16-core AMD Opteron 6274 (Interlagos) CPU.

The main bottleneck of this method is computation of the Schur complements of the stiffness matrices. This bottleneck is significantly reduced by using new PARDISO-SC sparse direct solver. This paper also presents the performance evaluation of SC computations for three-dimensional elasticity stiffness matrices.

We present the performance evaluation of the proposed approach using our implementation in the ESPRESO solver package.

Keywords: FETI solver · GPGPU · CUDA · Schur complement · ESPRESO

1 Introduction

The goal of this paper is to describe the acceleration of the Finite Element Tearing and Interconnection (FETI) method using Graphic Processing Units (GPU).

© Springer International Publishing Switzerland 2016
T. Kozubek et al. (Eds.): HPCSE 2015, LNCS 9611, pp. 85–100, 2016.
DOI: 10.1007/978-3-319-40361-8_6

The proposed method is based on our variant of FETI type domain decomposition method called Total FETI (TFETI) [2]. The original FETI method, also called the FETI-1, was originally developed for the numerical solution of large linear systems arising in linearized engineering problems by Farhat and Roux [1]. In FETI methods a body is decomposed into several non-overlapping subdomains and the continuity between the subdomains is enforced by Lagrange multipliers. Using the theory of duality, a smaller and relatively well-conditioned dual problem is derived and efficiently solved by a suitable variant of the conjugate gradient (CG) algorithm.

The original FETI algorithm, where only the favorable distribution of the spectrum of the dual Schur complement (SC) matrix [6] was considered, was efficient only for a small number of subdomains. So it was later extended by introducing a natural coarse problem [6,7], whose solution was implemented by auxiliary projectors so that the resulting algorithm became in a sense optimal [6,7]. Even if there are several efficient coarse problem parallelization strategies [8], the size limit is always present.

In the TFETI method [2], the Dirichlet boundary conditions are also enforced by Lagrange multipliers. Hence all subdomain stiffness matrices are singular with a priori known kernels which is a great advantage in the numerical solution. With known kernel basis we can effectively regularize the stiffness matrix [3] and use any standard Cholesky-type decomposition for nonsingular matrices. From the implementation point of view TFETI handles each subdomain in the same way. This significantly simplifies the implementation. The nonzero Dirichlet boundary conditions can be also implemented in a less complicated manner. More details regarding the efficiency and other aspects of the TFETI and other can be found in [2].

The stiffness matrices resulting from the three-dimensional elasticity problems are very sparse and they have to be treated in this fashion during the entire runtime of the solver. The operations on stiffness matrices done by FETI solvers therefore are (i) a factorization during the preprocessing stage —single call, and (ii) a backward and forward substitution (a solve routine) —called in every iteration. The second operation is the most time consuming and therefore this paper presents an approach that accelerates this part. For these two operations our ExaScale PaRallel FETI SOlver (ESPRESO) uses the PARDISO sparse direct solver [9–11].

In order to be able to efficiently utilize the potential of the GPU accelerators the solve routine of the sparse direct solver, which is naturally sequential for single right-hand side, has to be replaced by a more parallel operation. To achieve this we have modified the FETI algorithm so that it uses dense SC instead of sparse matrices. The SC is dense and its dimension is given by the size of a subdomain surface rather than its volume. By using the SC three sparse operations, two sparse matrix vector multiplications (SpMV) with gluing matrix $\mathbf{B}^s$ and one solve routine of the PARDISO, can be replaced by a single dense general matrix vector multiplication (GEMV) with the SC. The GEMV operation is well suited for GPU accelerators, as it offers coalesced memory access pattern and a high degree of parallelism.

1.1   Total FETI Method

The FETI-1 method [2] is based on the decomposition of the spatial domain into non-overlapping subdomains that are glued by Lagrange multipliers $\boldsymbol{\lambda}$, enforcing arising equality constraints by special projectors. The original FETI-1 method assumes that the boundary subdomains inherit the Dirichlet conditions from the original problem. This means physically that subdomains touching the Dirichlet boundary are fixed while the others remain floating; in linear algebra language this means that corresponding subdomain stiffness matrices are nonsingular and singular, respectively. The basic idea of our TFETI [2] is to keep all the subdomains floating and enforce the Dirichlet boundary conditions by means of a constraint matrix and Lagrange multipliers, similarly to the gluing conditions along subdomain interfaces. This simplifies implementation of the generalized inverse. The key point is that kernels of subdomain stiffness matrices are: (i) known a priori; (ii) have the same dimension; and (iii) can be formed without any computation from mesh data. Furthermore, each local stiffness matrix can be regularized cheaply, and the inverse of the resulting nonsingular matrix is at the same time a generalized inverse of the original singular one [3–5].

   Let N_p, N_d, N_n, N_c denote the primal dimension, the dual dimension, the null space dimension, and the number of cores available for our computation. Primal dimension means the number of all DOFs including those arising from duplication on the interfaces. Dual dimension is the total number of all equality constraints. Let us consider a partitioning of the global domain Ω into N_S subdomains $\Omega^s, s = 1, \ldots, N_S$ $(N_S \geq N_c)$. To each subdomain Ω^s there corresponds the subdomain stiffness matrix $\mathbf{K}^s$ and the subdomain nodal load vector $\mathbf{f}^s$. Matrix $\mathbf{R}^s$ shall be a matrix whose columns span the nullspace (kernel) of $\mathbf{K}^s$. Let $\mathbf{B}^s$ be a signed boolean matrix defining connectivity of the subdomain s with neighbor subdomains. It also enforces Dirichlet boundary conditions when TFETI is used. Using proposed notation the discretized equilibrium equation for s-th subdomain reads

$$\mathbf{K}^s \mathbf{u}^s = \mathbf{f}^s - (\mathbf{B}^s)^T \boldsymbol{\lambda} \tag{1}$$

where additional part in the RHS $(\mathbf{B}^s)^T \boldsymbol{\lambda}$ reflects the influence of neighboring subdomains trough selected LM defined by Boolean matrix $\mathbf{B}^s$. Objects from all subdomains can be collected:

$$
\begin{aligned}
\mathbf{K} &= \operatorname{diag}(\mathbf{K}^1, \ldots, \mathbf{K}^{N_S}) & &\in \mathbb{R}^{N_p \times N_p}, \\
\mathbf{R} &= \operatorname{diag}(\mathbf{R}^1, \ldots, \mathbf{R}^{N_S}) & &\in \mathbb{R}^{N_p \times N_n}, \\
\mathbf{B} &= [\mathbf{B}^1, \ldots, \mathbf{B}^{N_S}] & &\in \mathbb{R}^{N_d \times N_p}, \\
\mathbf{f} &= [(\mathbf{f}^1)^T, \ldots, (\mathbf{f}^{N_S})^T]^T & &\in \mathbb{R}^{N_p \times 1}, \\
\mathbf{u} &= [(\mathbf{u}^1)^T, \ldots, (\mathbf{u}^{N_S})^T]^T & &\in \mathbb{R}^{N_p \times 1}, \\
\boldsymbol{\lambda} & & &\in \mathbb{R}^{N_d \times 1},
\end{aligned} \tag{2}
$$

then global equilibrium equation is written as

$$\mathbf{K}\mathbf{u} = \mathbf{f} - \mathbf{B}^T \boldsymbol{\lambda} \tag{3}$$

which is completed with following compatibility condition

$$\sum_{i=1}^{N_s} \mathbf{B}^s \mathbf{u}^s = \mathbf{B}\mathbf{u} = \mathbf{o}. \tag{4}$$

Previous Eqs. (3) and (4) can be written together as

$$\mathbf{Ax} = \begin{pmatrix} \mathbf{K} \ \mathbf{B}^T \\ \mathbf{B} \ \mathbf{O} \end{pmatrix} \begin{pmatrix} \mathbf{u} \\ \boldsymbol{\lambda} \end{pmatrix} = \begin{pmatrix} \mathbf{f} \\ \mathbf{o} \end{pmatrix}. \tag{5}$$

The system (5) is reduced by elimination of primal variables $\mathbf{u}$. From Eq. (3) we can derive

$$\mathbf{u} = \mathbf{K}^+ \left(\mathbf{f} - \mathbf{B}^T \boldsymbol{\lambda} \right) + \mathbf{R}\boldsymbol{\alpha} \tag{6}$$

and combining with (4) we get

$$\mathbf{B}\mathbf{u} = \mathbf{B}\mathbf{K}^+ \mathbf{f} - \mathbf{B}\mathbf{K}^+ \mathbf{B}^T \boldsymbol{\lambda} + \mathbf{B}\mathbf{R}\boldsymbol{\alpha} = \mathbf{o}. \tag{7}$$

Note, that after elimination of $\mathbf{u}$ a new vector of unknowns $\boldsymbol{\alpha}$ appeared (amplitudes of rigid body motions); therefore an additional equation has to be added.

This additional equation follows immediately from the solvability of the first equation in (3), i.e.,

$$\mathbf{f} - \mathbf{B}^T \boldsymbol{\lambda} \in Image(\mathbf{K}) \tag{8}$$

which can be expressed equivalently using the matrix $\mathbf{R}$ as follows

$$\mathbf{R}^T \mathbf{K}\mathbf{u} = \mathbf{R}^T \mathbf{f} - \mathbf{R}^T \mathbf{B}^T \boldsymbol{\lambda} = \mathbf{o}. \tag{9}$$

Note that columns of $\mathbf{R}$ also span the kernel of $\mathbf{K}$, thus $\mathbf{R}^T \mathbf{K}\mathbf{u} = \mathbf{o}$. Now both (7) and (9) written together,

$$\begin{pmatrix} \mathbf{B}\mathbf{K}^+ \mathbf{B}^T \ -\mathbf{B}\mathbf{R} \\ -\mathbf{R}^T \mathbf{B}^T \ \mathbf{O} \end{pmatrix} \begin{pmatrix} \boldsymbol{\lambda} \\ \boldsymbol{\alpha} \end{pmatrix} = \begin{pmatrix} \mathbf{B}\mathbf{K}^+ \mathbf{f} \\ -\mathbf{R}^T \mathbf{f}^T \end{pmatrix}, \tag{10}$$

define the reduced system. Using following notations,

$$\begin{aligned} \mathbf{F} &= \mathbf{B}\mathbf{K}^+ \mathbf{B}^T = \sum_{s=1}^{N_s} \mathbf{F}^s = \sum_{s=1}^{N_s} \mathbf{B}^s (\mathbf{K}^s)^+ (\mathbf{B}^s)^T, \\ \mathbf{G} &= -\mathbf{R}^T \mathbf{B}^T, \\ \mathbf{d} &= \mathbf{B}\mathbf{K}^+ \mathbf{f} = \sum_{s=1}^{N_s} \mathbf{d}^s = \sum_{s=1}^{N_s} \mathbf{B}^s (\mathbf{K}^s)^+ \mathbf{f}^s, \\ \mathbf{e} &= -\mathbf{R}^T \mathbf{f}, \end{aligned} \tag{11}$$

it can be rewritten as

$$\begin{pmatrix} \mathbf{F} \ \mathbf{G}^T \\ \mathbf{G} \ \mathbf{O} \end{pmatrix} \begin{pmatrix} \boldsymbol{\lambda} \\ \boldsymbol{\alpha} \end{pmatrix} = \begin{pmatrix} \mathbf{d} \\ \mathbf{e} \end{pmatrix}. \tag{12}$$

The obtained matrix

$$\mathbf{F} = -\mathbf{S} = \mathbf{B}\mathbf{K}^+ \mathbf{B}^T \tag{13}$$

is actually the negative SC of the system (5). Compared to the original KKT system (5) the new one (12) has smaller dimension and it is also better conditioned.

The equality constraints $\mathbf{G}\boldsymbol{\lambda} = \mathbf{e}$ in (12) can be homogenized to $\mathbf{G}\bar{\boldsymbol{\lambda}} = \mathbf{o}$ by splitting $\boldsymbol{\lambda}$ into $\bar{\boldsymbol{\lambda}} + \widetilde{\boldsymbol{\lambda}}$, where $\widetilde{\boldsymbol{\lambda}}$ satisfies $\mathbf{G}\widetilde{\boldsymbol{\lambda}} = \mathbf{e}$. This implies $\bar{\boldsymbol{\lambda}} \in \mathrm{Ker}\,\mathbf{G}$. The vector $\widetilde{\boldsymbol{\lambda}}$ can be chosen as the least squares solution of the equality constraints, i.e. $\widetilde{\boldsymbol{\lambda}} = \mathbf{G}^T(\mathbf{G}\mathbf{G}^T)^{-1}\mathbf{e}$. We substitute $\boldsymbol{\lambda} = \bar{\boldsymbol{\lambda}} + \widetilde{\boldsymbol{\lambda}}$, minimize over $\bar{\boldsymbol{\lambda}}$ (terms without $\bar{\boldsymbol{\lambda}}$ can be omitted) and add $\widetilde{\boldsymbol{\lambda}}$ to $\bar{\boldsymbol{\lambda}}$.

Finally, equality constraints $\mathbf{G}\bar{\boldsymbol{\lambda}} = \mathbf{o}$ can be enforced by the orthogonal projector $\mathbf{P} = \mathbf{I} - \mathbf{Q}$ onto the null space of $\mathbf{G}$, where $\mathbf{Q} = \mathbf{G}^T(\mathbf{G}\mathbf{G}^T)^{-1}\mathbf{G}$ is the orthogonal projector onto the image space of $\mathbf{G}^T$ (i.e., $\mathrm{Im}\,\mathbf{Q} = \mathrm{Im}\,\mathbf{G}^T$ and $\mathrm{Im}\,\mathbf{P} = \mathrm{Ker}\,\mathbf{G}$). The final problem reads

$$\mathbf{P}\mathbf{F}\bar{\boldsymbol{\lambda}} = \mathbf{P}\widetilde{\mathbf{d}}, \tag{14}$$

where $\widetilde{\mathbf{d}} = \mathbf{d} - \mathbf{F}\widetilde{\boldsymbol{\lambda}}$. Note that we call the action of $(\mathbf{G}\mathbf{G}^T)^{-1}$ the *coarse problem* of FETI. Problem (14) can be solved with an arbitrary iterative linear system solver. The conjugate gradient method is a good choice thanks to the classical estimate by Farhat et al. [6] of the spectral condition number

$$\kappa(\mathbf{P}\mathbf{F}\mathbf{P}|\mathrm{Im}\mathbf{P}) \leq C\frac{H}{h}. \tag{15}$$

2 Stiffness Matrix Preprocessing and Schur Complement Computation

The FETI solver processing can be divided into two stages: (i) the preprocessing and (ii) the solver runtime. During the preprocessing stage the most time consuming tasks include assembling the distributed inverse matrix of coarse problem $(\mathbf{G}\mathbf{G}^T)^{-1}$ and factorization of the subdomain stiffness matrices $\mathbf{K}^s$. While the coarse problem processing time is mainly given by the number of subdomains, the $\mathbf{K}^s$ factorization times are given by the subdomain sizes. The solver runtime stage executes the CG iterative solver and its execution time is therefore given by the number of iterations and the iteration time. If the iteration time is evaluated then the most time consuming part is executing the solve routine (forward and backward substitution) of the sparse direct solver using the Cholesky decomposition of the $\mathbf{K}^s$. This operation takes up to 95 % of the iteration time. The main focus of this paper is to describe a method to accelerate this operation using GPU accelerators.

The sparse data structures cannot take full advantage of modern processing hardware such as GPU accelerators which is equipped with Single Instruction Multiple Data (SIMD) units. To reach the full potential of a GPU architecture a dense representation of the data that were originally sparse is the key task. In case of FETI solvers the data that has to be converted to dense structure is the Cholesky decomposition of stiffness matrices $\mathbf{K}^s$.

In the FETI solver prior to calling the solve routine, the vector of Langrange multipliers (the dual variables $\bar{\boldsymbol{\lambda}}$) has to be converted to primal variables using gluing matrix $\mathbf{B}^T$. The result of the solve routine (primal variables) have to be again converted into dual variables using $\mathbf{B}$. This operation in the algorithm of the projected preconditioned conjugate gradient method appears during the initialization stage, and then once in each iteration (in Algorithm 1 the appearances marked by the symbol $\star$). This step

> **Data:** set $\bar{\boldsymbol{\lambda}} = \mathbf{P}\bar{\boldsymbol{\lambda}}_0,\ \bar{\boldsymbol{\lambda}}_0 \in \mathbb{R}^n,\ \varepsilon > 0,\ i_{max} > 0$
> $\mathbf{M} \approx (\mathbf{PFP})^{-1}$
> $\mathbf{g} = \mathbf{F}\bar{\boldsymbol{\lambda}} - \tilde{\mathbf{d}}\ (\star)$
> $\mathbf{w} = \mathbf{PMPg}$
> **for** $i = 0, 1, \cdots, i_{max}$ **do**
> $\quad$ $\mathbf{v} = \mathbf{Fp}\ (\star)$
> $\quad$ $\mathbf{w}_{prev} = \mathbf{w}$
> $\quad$ $\mathbf{w} = \mathbf{PMPg}$
> $\quad$ $\rho = -(\mathbf{g}, \mathbf{w})/(\mathbf{p}, \mathbf{v})$
> $\quad$ $\bar{\boldsymbol{\lambda}} = \bar{\boldsymbol{\lambda}} + \mathbf{p}\rho$
> $\quad$ $\mathbf{g}_{prev} = \mathbf{g}$
> $\quad$ $\mathbf{g} = \mathbf{g} + \mathbf{v}\rho$
> $\quad$ **if** $\sqrt{(\mathbf{g}, \mathbf{Pg})} < \varepsilon$ **then**
> $\quad$ | $\quad$ **break**
> $\quad$ **end**
> $\quad$ $\beta = (\mathbf{g}, \mathbf{w})/(\mathbf{g}_{prev}, \mathbf{w}_{prev})$
> $\quad$ $\mathbf{p} = \mathbf{w} + \mathbf{p}\beta$
> **end**
> $\boldsymbol{\lambda} = \tilde{\boldsymbol{\lambda}} + \bar{\boldsymbol{\lambda}}$

Algorithm 1. The projected preconditioned conjugate gradient method.

$$\mathbf{v} = \sum_{s=1}^{N_s} \mathbf{B}^s (\mathbf{K}^s)^+ (\mathbf{B}^s)^T \mathbf{p}, \tag{16}$$

which in regular CG is equal to matrix-vector multiplication with the system matrix, contains two calls of sparse matrix vector multiplications of sparse BLAS (SpMV with $\mathbf{B}^s$) and one call of PARDISO solve routine (action of generalized inverse $(\mathbf{K})^+$). Instead of executing these three operations on sparse matrices, we can directly assemble the sth contribution $\mathbf{F}^s$ to the global FETI operator $\mathbf{F}$ from

$$\mathbf{A}^s = \begin{pmatrix} \mathbf{K}^s & (\mathbf{B}^s)^T \\ \mathbf{B}^s & 0 \end{pmatrix}. \tag{17}$$

We evaluated two approaches how to assemble the matrix $\mathbf{F}^s$. The first approach is by solving the systems $\mathbf{K}^s \mathbf{X}^s = (\mathbf{B}^s)^T$ and then calculating $\mathbf{F}^s = \mathbf{B}^s \mathbf{X}^s$.

The second and more efficient approach is using the incomplete factorization method implemented in PARDISO-SC applied to the system in (17) where $\mathbf{F}^s$ will sit in the block (2,2) if the factorization is stopped after the elimination of the (1,1) block of the system in (17). This method is approximately 3.3 to 4.5

times faster depending on system size; see Table 2. The same table also shows that assembling $\mathbf{F}^s$ is slower than factorization of the matrix $\mathbf{K}^s$ by 13.9 to 18.6 times. It is therefore important to evaluate the trade off between preprocessing time and solver runtime.

3 GPU Acceleration of the FETI Iterative Solver

In general, one of the most limiting factors of GPU acceleration is data transfers between the CPU and GPU main memory over the PCI-Express bus. In our approach the calculation of the Schur complements $\mathbf{F}^s$ is approximately 20 to 60 times slower than the data transfers to GPU if data is transferred only once during the preprocessing stage. However it is not possible to maintain high performance if solver needs to transfer $\mathbf{F}^s$ matrices back and forth between CPU and GPU memory in every iteration. Therefore the size of the GPU global memory is limiting the maximum problem size that can be processed per accelerator.

In case of Tesla K20X with 6 GB of memory, it is able to solve problem of size approximately 0.5 million of DOFs if the $\mathbf{F}^s$ matrices are stored as general dense matrices. This size depends on the surface-to-volume ratio of the subdomains and hence the partitioning. This particular case is for cubic subdomains. For decomposition into different number of cubic subdomains see Table 2. But, since the $\mathbf{K}^s$ is symmetric, the $\mathbf{F}^s$ is also symmetric and it can be stored using the packed format. In this case the problem size solved by one GPU is two times larger.

Since all the $\mathbf{F}^s$ matrices are transferred during the preprocessing only input and output vectors are transferred per iteration. To be able to hide these transfers behind the execution of the GEMV kernel on GPU each subdomain uses its own CUDA streams. Streams are initialized during the preprocessing prior to transfer the $\mathbf{F}^s$ to GPU. Then in every iteration the ith stream transfers the input vector, executes cuBLAS GEMV kernel, and transfers output vector back to CPU memory, where i is the subdomain index. All streams are eventually synchronized to ensure that all output vectors are successfully transferred back to the CPU memory. Then the CG algorithm can continue.

CuBLAS provides three matrix-vector multiplication routines that can be used by our approach: (1) GEMV - general matrix-vector multiplication with storage requirements n^2, (2) SYMV - symmetric matrix-vector multiplication with storage requirements n^2 and (3) SPMV - symmetric packed matrix-vector multiplication with storage requirements $n(n + 1)/2$. While the last routine is optimal due to reduced memory usage its performance is significantly lower. We have evaluated the performance of all three routines using cuBLAS version 6.5. The fastest kernel is GEMV. The SYMV kernel performance is lower by 5 % (it is based on implementation in KBLAS). The slowest performance was achieved by SPMV with packed format which is up to 3 times slower then GEMV.

Therefore the optimal choice is to use the SYMV kernel and let two subdomains share the memory allocation of size $n(n + 2)$, where the first subdomain uses the upper triangular part of the allocation and the second subdomain uses the lower part. This way the GPU memory usage is optimal, and performance penalty is only 5 %. This implies a problem with implementation as two subdomains have to share single memory allocation and the sizes of these two subdomains have to be similar to achieve optimal memory usage.

The proposed approach supports the use of preconditioners in the identical way that the original Total FETI method.

4   Acceleration of Computation of Schur Complement by PARDISO-SC

Let us solve the system introduced in (5) in the itemized form

$$
\begin{pmatrix} \mathbf{K}^1 & & & (\mathbf{B}^1)^T \\ & \ddots & & \vdots \\ & & \mathbf{K}^{N_S} & (\mathbf{B}^{N_S})^T \\ \mathbf{B}^1 & \cdots & \mathbf{B}^{N_S} & \mathbf{O} \end{pmatrix} \begin{pmatrix} \mathbf{u}^1 \\ \vdots \\ \mathbf{u}^{N_S} \\ \lambda \end{pmatrix} = \begin{pmatrix} \mathbf{f}^1 \\ \vdots \\ \mathbf{f}^{N_S} \\ \mathbf{o} \end{pmatrix}
\tag{18}
$$

in which we assume all diagonal blocks $\mathbf{K}^s$ are nonsingular (to utilize PARDISO also for cases where matrices $\mathbf{K}^s$ are singular, a special regularization technique described in [3] can be used). Solution of such a linear system is given by Schur complement method. First, the SC $\mathbf{S}$ is computed,

$$
\mathbf{S} = -\mathbf{F} = \sum_{s=1}^{N_S} \mathbf{S}^s,
\tag{19}
$$

where

$$
\mathbf{S}^s = -\mathbf{B}^s (\mathbf{K}^s)^{-1} (\mathbf{B}^s)^T
\tag{20}
$$

and then the first-stage part of the solution of the system (5) solved with FETI operator $\mathbf{F}$ instead of $\mathbf{S}$ is obtained by solving

$$
\mathbf{F}\lambda = \sum_{s=1}^{N} \mathbf{B}^s (\mathbf{K}^s)^{-1} \mathbf{f}^s.
\tag{21}
$$

Finally, the second-stage part of the solution can be obtained for all

$$
\mathbf{K}^s \mathbf{u}^s = \mathbf{f}^s - (\mathbf{B}^s)^T \lambda.
\tag{22}
$$

A quick look at (20)–(22) reveals great scope for parallelism. More specifically, the computation of the contributions $\mathbf{B}^s (\mathbf{K}^s)^{-1} (\mathbf{B}^s)^T$ to the SC, the evaluation of the residual in (21), and the solutions $\mathbf{u}^i$, $s = 1, \ldots, N_S$, can be performed independently.

As we previously mentioned, the data are sparse and unstructured in the case of our application; therefore the computation of $\mathbf{B}^s (\mathbf{K}^s)^{-1} (\mathbf{B}^s)^T$ needs to rely on *sparse* linear algebra kernels. Previously, we used off-the-shelf sparse linear solvers such as PARDISO to first factorize $\mathbf{K}^s$ as $\mathbf{L}^s\mathbf{D}^s(\mathbf{L}^s)^T$, then perform triangular solves with the factors of $\mathbf{K}^s$ for each nonzero column of $(\mathbf{B}^s)^T$, i.e., compute $(\mathbf{K}^s)^{-1}(\mathbf{B}^s)^T$, and, finally, multiply the result from the left with $\mathbf{B}^s$. This approach has two important drawbacks in multicore environments: (i) the triangular solves with $\mathbf{L}^s$ and $(\mathbf{L}^s)^T$ do not scale well with the number of cores [13], being memory bound, and (ii) the sparsity of the columns of $(\mathbf{B}^s)^T$ may not be exploited by some sparse direct linear solvers when solving $\mathbf{L}^s\mathbf{X} = (\mathbf{B}^s)^T$. This feature has been for instance added to the recent version of the open-source MUMPS direct solver.

To address these limitations the computations were revisited and new approach was proposed in [12]. This new approach computes $\mathbf{B}^s(\mathbf{K}^s)^{-1}(\mathbf{B}^s)^T$ from a partial sparse Bunch–Kaufman factorization of the augmented matrix in (17). More specifically, the factorization of $\mathbf{A}^s$ is stopped after pivoting reaches the last diagonal entry of $\mathbf{K}^s$. At this point $-\mathbf{B}^s(\mathbf{K}^s)^{-1}(\mathbf{B}^s)^T$ is computed and resides in the $(2,2)$ block of $\mathbf{A}^s$.

Traditionally, the factorizing phase has generally required the greatest portion of the total execution time. It typically involves the majority of the floating-point operations, and it is also computation bound. However, in this application, we have to compute $\mathbf{B}^s(\mathbf{K}^s)^{-1}(\mathbf{B}^s)^T$, and memory traffic is the limiting factor. In exploiting the sparsity not only in $\mathbf{K}^s$, but also in $(\mathbf{B}^s)^T$, we (i) significantly reduce the number of floating point operations and (ii) can use in-memory sparse matrix compression technique to reduce the memory traffic on multicore architectures. In the numerical section we refer to this compression techniques as PARDISO-SC, whereas PARDISO is the uncompressed triangular solve based on $\mathbf{K}^s$. As a result, the approach in PARDISO-SC is much better suited for multicore parallelization than the triangular solves and, consequently, the speedup over the previous approach is quite considerable, as we will show in numerical experiments.

4.1 Parallelization Techniques in ESPRESO Solver

The ESPRESO is a FETI-based sparse iterative solver implemented in C++. The solver is developed to support modern heterogeneous accelerators such as Intel Xeon Phi or Nvidia GPU. Due to this fact, the CPU version uses the Intel MKL library and in addition the cuBLAS library is used for GPU acceleration. A significant part of the development effort was devoted to writing a C++ wrapper for (1) the selected sparse and dense BLAS routines of the Intel MKL library and (2) the sparse direct solvers. As of now PARDISO and PARDISO-SC are supported. By simple modification of the wrapper support for additional direct solvers can be added. This is an ongoing work.

The hybrid parallelization inside the ESPRESO is designed to fit two-level decomposition used by the Hybrid FETI Method. This method decomposes the problem into clusters (the first level), then the clusters are further decomposed into subdomains (the second level). In case of the TFETI method the problem is directly decomposed into subdomains and predefined number of subdomains is assigned to a particular compute node. In this paper we focus on the TFETI method only.

In ESPRESO this decomposition is mapped to a parallel hardware in the following way: (level 1) groups of subdomains are mapped to compute nodes, communication between nodes is done using message passing—MPI; and (level 2) subdomains inside nodes are mapped to CPU cores using Cilk++ shared memory model. ESPRESO supports the oversubscription of the CPU cores which means that multiple subdomains are processed by a single CPU core. In a case of the GPU acceleration all subdomains of one node are processed by a single GPU. Then using CUDA streams multiple subdomains can be processed in parallel. Support of multiple GPU accelerators per node is also supported. More about GPU implementation is described in Sect. 3.

There are two major parts of the solver that affect its parallel performance and scalability: (i) the communication layer (described in this section) and (ii) the inter node processing routines for shared memory. The second part can be further divided into CPU processing routines and highly computationally intensive routines suitable for offloading to the accelerator.

The first part deals with optimization of the communication overhead caused mainly by multiplication with gluing matrices $\mathbf{B}$, application of the projector (includes multiplication with matrix $\mathbf{G}$ and the application of the coarse problem), and global reduction operation in the CG solver. The distributed memory parallelization is done using MPI 3.0 standard because the communication hiding techniques for an iterative CG solver require the nonblocking MPI collective operations.

The following communication avoiding and hiding techniques for the main CG iterative solver are used: (i) the pipelined conjugate gradient (PipeCG) iterative solver—hides communication cost of the global dot products in CG behind the local matrix vector multiplications; (ii) the coarse problem solver using a distributed inverse matrix—merges two global communication operations (Gather and Scatter) into one (AllGather) and parallelizes the coarse problem processing; and (3) the optimized version of global gluing matrix multiplication ($\mathbf{B}^s$)—implemented as a stencil communication which is fully scalable. The potential of the ESPRESO communication layer is shown in Fig. 1.

5   Results

The performance evaluation has been carried out using the Titan machine installed in the Oak Ridge National Lab. Each compute node contains one 16-core 2.2 GHz AMD Opteron 6274 (Interlagos) processor, 32 GB of RAM and NVIDIA Tesla K20X accelerator (GPU) with 6 GB of DDR5 memory.

For the evaluation a synthetic 3D cube benchmark is used. The size of the problem is limited by the size of the GPU memory. Tesla K20X has 6 GB of RAM and it can fit problems of size approximately 0.5 million of unknowns if stored as general matrices. Tests were executed for decomposition into 48, 64, 80, 100, 150 and 216 cubical subdomains. For more details regarding the decomposition please see the Table 2.

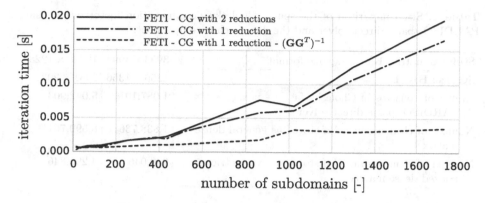

Fig. 1. Performance evaluation of optimization techniques introduced into the communication layer of ESPRESO. (i) CG algorithm with 2 reductions—is the standard version of the CG algorithm. (ii) CG algorithm with 1 reduction —is based on the preconditioned pipelined CG algorithm, where projector is used in place of preconditioner (this algorithm is ready to use nonblocking global reduction for further performance improvements (comes in Intel MPI 5.0)), and (iii) GGTINV—parallelizes the solve of the coarse problem plus merges two Gather and Scatter global operations into single AllGather. Please note that very small subdomains are used (648 DOFs)—small subdomain size is chosen to identify all communication bottlenecks of the solver .

5.1   Calculation of the Schur Complement Using Sparse Direct Solvers

In this section an evaluation of the two methods used to compute the Schur complements $\mathbf{F}^s$ for FETI solver is presented. Two metrics are considered· (i) memory requirements and (ii) processing time to compute the SCs. All results are compared to the original method which uses sparse Cholesky decomposition. This evaluation has been carried out for a cubical subdomains.

In Table 1 the size of the dense $\mathbf{F}^s$ structure is compared to number of nonzero elements in the sparse Cholesky decomposition. This represents the memory requirements of both approaches. It can be observed that if the $\mathbf{F}^s$ is stored in packed format (only upper or lower triangle with the diagonal is kept in memory) its size is lower that number of non zero values in the Cholesky decomposition. If we take into account that the sparse representation of data requires more than 1 integer value (4 bytes) and 1 double value (8 bytes) per element we can see that SC method is more efficient in terms of memory usage.

Table 2 compares the stiffness matrix preprocessing times of all three approaches on a problem of size approximately 0.5 million DOFs. The problem is decomposed into various number of subdomains and it can be observed that as the size of the subdomain is getting smaller the calculation of the SC is getting more efficient when compared to the factorization time. See the second values in the PARDISO-SC column which describe how many times calculation of SC is slower than factorization. The advantage of using PARDISO-SC is obvious when processing time is compared to last column of the table which contains

Table 1. Size comparison of the sparse Cholesky decomposition/factorization using PARDISO sparse direct solver and the dense Schur complement.

Stiffness matrix ($\mathbf{K}$) size-sparse format	3993×3993	12288×12288
SC size: $\mathbf{BK^+B}^T$	1356×1356	2931×2931
Nmber of nonzeros in Cholesky factor - using PARDISO sparse direct solver	1,087,101	5,042,903
Number of elements in SC if stored as general dense matrix	1,838,736	8,590,761
Number of elements in SC if stored as symmetric packed dense matrix	920,046	4,296,846

Table 2. Performance evaluation of the stiffness matrix preprocessing time on a single node of the Oak Ridge National Lab machine called Titan. The second values in PARDISO-SC column describe how many times it is slower than factorization. The second values in PARDISO column describe how many times it is slower than PARDISO-SC in calculation of the Schur complement.

Problem size [DOFs]	Number of subdomains [-]	Subdomain size [DOFs]	Preprocessing time factorization	Preprocessing time - Schur complement PARDISO-SC	Preprocessing time - Schur complement PARDISO
513 498	48	12288	4.1 s	76.5 s ; 18.6	256.2 s ; 3.3
419 121	48	10125	3.1 s	56.4 s ; 18.4	233.1 s ; 4.1
446 631	64	8232	2.9 s	49.6 s ; 16.9	210.3 s ; 4.2
439 383	80	6591	2.6 s	42.0 s ; 16.1	190.0 s ; 4.5
423 360	100	5184	2.4 s	34.0 s ; 13.9	144.6 s ; 4.3
475 983	150	3993	2.4 s	34.9 s ; 14.4	134.8 s ; 3.9
499 125	216	3000	2.2 s	30.6 s ; 13.7	112.9 s ; 3.7

computational times of SC using PARDISO. We can see that PARDISO-SC is 3.3 to 4.5 times faster.

5.2   GPU Acceleration of the Conjugate Gradient Solver in Total FETI

The previous section described the additional work that needs to be done in the preprocessing stage. In this section we describe the main performance gain by using the SC in FETI, which is the acceleration of the CG solver.

The results are shown in Table 3. The problem of the same size (0.5 millions of DOFs) and the decomposition into the same number of subdomains as in the previous section is used. It can be seen that GPU acceleration is more efficient for larger subdomains, speedup of 4.1 for subdomain of size 12 288 DOFs and it is getting less efficient for smaller subdomains, speedup of 1.8 for subdomain of size 3000 DOFs. This is an opposite behavior to the preprocessing stage where calculation of the Schur complement is more costly for larger subdomains.

The execution time values in Table 3 show the solver runtime for 500 iterations. Based on our measurements this is the sweet spot where GPU acceleration for this hardware configuration becomes more efficient than CPU version.

Table 3. Iterative solver execution time for 500 iterations running on a single Tesla K20X GPU accelerator. Table compares solve routine which includes forward and backward substitution of the PARDISO solver and GEMV (general matrix vector multiplication) routine from the cuBLAS GPU accelerated library.

Problem size [DOFs]	Number of subdomains [-]	Subdomain size [DOFs]	Solver runtime on GPU [s]	Solver runtime on CPU [s]	Speedup by GPU [-]
513 498	48	12288	27.76	113.63	4.1
419 121	48	10125	22.99	87.68	3.8
446 631	64	8232	24.46	90.74	3.7
439 383	80	6591	25.33	81.58	3.2
423 360	100	5184	25.94	74.66	2.9
475 983	150	3993	33.03	78.03	2.4
499 125	216	3000	40.29	74.12	1.8

5.3 Overall Performance for Different Classes of Problems in 3D Linear Elasticity

The overall performance which includes preprocessing and solver runtime is shown in Table 4. In the current version the ESPRESO library can generate and solve linear elasticity problems in three dimensions. If we focus on this group of problems, we can evaluate the usability of the approach presented in this paper for three scenarios. The first scenario is a simple linear elasticity problem. In this case the number of iterations is usually less than 100. It is clear that for this types of problems the preprocessing time is dominant and presented approach including GPU acceleration does not bring any advantage. The second scenario is for group of ill-conditioned linear elasticity problems. These problems approximately converge in several hundreds of iterations. This is the border line situation where assembling Schur complements becomes useful and overall execution time is reduced. The gain for this type of problems is not dramatic. The last group are transient and nonlinear problems with constant tangent stiffness matrix. To solve these problems solver has to perform tens to hundreds of iterations in every time or non-linear solver step. This generates thousands of iterations and the presented approach has a potential to run up to 4.1 times faster.

Table 4. The overall performance evaluation of the GPU acceleration of the Total FETI solver in ESPRESO which includes both preprocessing and solver runtime. Three scenarios with 100, 500 and 2000 iterations are evaluated. The table shows that the sweet spot, where the GPU becomes more efficient is approximately around 500 iterations and more.

Decomposition num. of subdom.; subdom. size [DOFs]	Preproc. time [s] factorization; Schur compl.	100 iterations CPU[s]; GPU[s]; speedup[-]	500 iterations CPU[s]; GPU[s]; speedup[-]	2000 iterations CPU[s]; GPU[s]; speedup[-]
48; 12288	4.1; 76.5	26.8; 82.1; **0.3**	117.7; 104.3; **1.1**	458.6; 187.6; **2.4**
48; 10125	3.1; 56.4	20.6; 61.0; **0.3**	90.7; 79.3; **1.1**	353.8; 148.3; **2.4**
64; 8232	2.9; 49.6	21.1; 54.5; **0.4**	93.7; 74.1; **1.3**	365.9; 147.5; **2.5**
80; 6591	2.6; 42.0	18.9; 47.1; **0.4**	84.2; 67.3; **1.3**	328.9; 143.3; **2.3**
100; 5184	2.4; 34.0	17.4; 39.2; **0.4**	77.1; 60.0; **1.3**	301.1; 137.8; **2.2**
150; 3993	2.4; 34.9	18.0; 41.5; **0.4**	80.4; 67.9; **1.2**	314.5; 167.0; **1.9**
216; 3000	2.2; 30.6	17.1; 38.7; **0.4**	76.3; 70.9; **1.1**	298.7; 191.8; **1.6**

6   Conclusions

Using the Schur complement opens the possibility to accelerate FETI algorithms by modern highly parallel hardware such as GPU accelerators. However this approach requires more expensive preprocessing in form of calculation of the Schur complement than original approach used in Total FETI. The preprocessing is between 13.7 to 18.6 times slower for the hardware configuration of the Titan supercomputer. This translates into penalty of 28.4 to 72.4 s for problems of size approximately 0.5 millions of degrees of freedom. Problems of this size fully utilize the 6 GB of Tesla K20X memory.

This paper presents the proof of concept, that manycore accelerators such as GPU or Intel Xeon Phi are valid hardware technology to make FETI solvers run faster. The GPU implementation is presented here, but the version for Xeon Phi is under active development. The problem evaluated in this paper, 0.5 million of unknowns decomposed into 216 domains, is rather small in order to fit single GPU with 6 GB of memory. This means that the coarse problem is not causing any performance bottleneck in this case. The evaluation of the coarse problem processing for Total FETI in ESPRESO is presented in [14]. We are currently working on extending the ESPRESO solver to support multiple accelerators per node.

Because the computation of the Schur complement for large subdomains becomes very costly it is optimal to decompose the problem into smaller subdomains. The subdomains of sizes 3000 to 12000 DOFs as presented in this paper are reasonable setting for current hardware architectures. Solving extremely large problems decomposed into small subdomains introduces very large coarse problem in case of Total FETI method. This becomes the main bottleneck for large problems. It is also the reason why Total FETI cannot be efficiently used to solve problems with several billions of unknowns.

The solution to this problem is to use multilevel decomposition. This method reduces the coarse problem size by grouping subdomains into clusters. Then the global coarse problem size is defined by a number of clusters and not by a number of subdomains. We are currently working on the implementation of a Hybrid FETI method into ESPRESO together with Schur complement approach.

Acknowledgment. This work was supported by The Ministry of Education, Youth and Sports from the National Programme of Sustainability (NPU II) project IT4Innovations excellence in science - LQ1602 and from the Large Infrastructures for Research, Experimental Development and Innovations project IT4Innovations National Supercomputing Center LM2015070; and by the EXA2CT project funded from the EUs Seventh Framework Programme (FP7/2007–2013) under grant agreement No. 610741.

References

1. Farhat, C., Roux, F.-X.: An unconventional domain decomposition method for an efficient parallel solution of large-scale finite element systems. SIAM J. Sci. Stat. Comput. **13**, 379–396 (1992)
2. Dostál, Z., Horák, D., Kučera, R.: Total FETI - an easier implementable variant of the FETI method for numerical solution of elliptic PDE. Commun. Numer. Methods Eng. **22**(12), 1155–1162 (2006)
3. Brzobohatý, T., Dostál, Z., Kozubek, T., Kovář, P., Markopoulos, A.: Cholesky decomposition with fixing nodes to stable computation of a generalized inverse of the stiffness matrix of a floating structure. Int. J. Numer. Methods Eng. **88**(5), 493–509 (2011). doi:10.1002/nme.3187
4. Dostál, Z., Kozubek, T., Markopoulos, A., Menšík, M.: Cholesky decomposition of a positive semidefinite matrix with known kernel. Appl. Math. Comput. **217**(13), 6067–6077 (2011). doi:10.1016/j.amc.2010.12.069
5. Kučera, R., Kozubek, T., Markopoulos, A.: On large-scale generalized inverses in solving two-by-two block linear systems. Linear Algebra Appl. **438**(7), 3011–3029 (2013)
6. Farhat, C., Mandel, J., Roux, F.-X.: Optimal convergence properties of the FETI domain decomposition method. Comput. Methods Appl. Mech. Eng. **115**, 365–385 (1994)
7. Roux, F.-X., Farhat, C.: Parallel implementation of direct solution strategies for the coarse grid solvers in 2-level FETI method. Contemp. Math. **218**, 158–173 (1998)
8. Kozubek, T., Vondrák, V., Menšík, M., Horák, D., Dostál, Z., Hapla, V., Kabelikova, P., Cermak, M.: Total FETI domain decomposition method and its massively parallel implementation. Adv. Eng. Softw. **60**, 14–22 (2013)
9. Kuzmin, A., Luisier, M., Schenk, O.: Fast methods for computing selected elements of the green's function in massively parallel nanoelectronic device simulations. In: Wolf, F., Mohr, B., an Mey, D. (eds.) Euro-Par 2013. LNCS, vol. 8097, pp. 533–544. Springer, Heidelberg (2013)
10. Schenk, O., Bollhöfer, M., Römer, R.: On large-scale diagonalization techniques for the Anderson model of localization. Featured SIGEST paper in the SIAM Review selected "on the basis of its exceptional interest to the entire SIAM community". SIAM Rev. **50**, 91–112 (2008)

11. Schenk, O., Wächter, A., Hagemann, M.: Matching-based preprocessing algorithms to the solution of saddle-point problems in large-scale nonconvex interior-point optimization. J. Comput. Optim. Appl. **36**(2–3), 321–341 (2007). doi:10.1007/s10589-006-9003-y
12. Petra, C., Schenk, O., Lubin, M., Gänter, K.: An augmented incomplete factorization approach for computing the Schur complement in stochastic optimization. SIAM J. Sci. Comput. **36**(2), C139–C162 (2014). doi:10.1137/130908737
13. Hogg, J.D., Scott, J.A.: A note on the solve phase of a multicore solver, SFTC Rutherford Appleton Laboratory, Technical report, Science and Technology Facilities Council, June 2010
14. Říha, L., Brzobohatý, T., Markopoulos, A.: Highly scalable FETI methods in ESPRESO. In: Ivnyi, P., Toppin, B.H.V. (eds.) Proceedings of the Fourth International Conference on Parallel, Distributed, Grid, Cloud Computing for Engineering, Civil-Comp Press, Stirlingshire, UK, Paper 17 (2015). doi:10.4203/ccp.107.17

Solving Contact Mechanics Problems
with PERMON

Vaclav Hapla[1,2(✉)], David Horak[1,2], Lukas Pospisil[1,2], Martin Cermak[1],
Alena Vasatova[1,2], and Radim Sojka[1,2]

[1] IT4Innovations National Supercomputing Center,
VSB - Technical University of Ostrava, Ostrava, Czech Republic
vaclav.hapla@vsb.cz
[2] Department of Applied Mathematics,
VSB - Technical University of Ostrava, Ostrava, Czech Republic

Abstract. PERMON makes use of theoretical results in quadratic programming algorithms and domain decomposition methods. It is built on top of the PETSc framework for numerical computations. This paper describes its fundamental packages and shows their applications. We focus here on contact problems of mechanics decomposed by means of a FETI-type non-overlapping domain decomposition method. These problems lead to inequality constrained quadratic programming problems that can be solved by our PermonQP package.

1 Introduction

We shall present our new software called PERMON (Parallel, Efficient, Robust, Modular, Object-oriented, Numerical) toolbox [17] and show its capabilities. PERMON extends PETSc [3] with support for quadratic programming (QP) and non-overlapping domain decomposition methods (DDM), namely of the FETI (Finite Element Tearing and Interconnecting) [5,12,13,24] type.

This paper presents the process of solving contact problems using PERMON (Sect. 3). We consider three model problems: two scalar contact problems of two membranes (coercive and semicoercive) and a contact problem of a 3D linear elastic cube with an obstacle (Sect. 2). The mesh is "teared" into subdomains and each of them is discretized separately and sequentially with the FEM (Finite Element Method). This decomposition and discretization is implemented by the PermonMembrane and PermonCube (Sect. 4) packages. The subdomain problems are then "interconnected" by means of FETI using PermonFLLOP (Sect. 5). Finally, the resulting QP problem is solved by the PermonQP module (Sect. 6). It contains implementations of specific algorithms for bound and equality constrained problems, particularly the SMALBE and MPRGP algorithms (Sects. 7 and 8).

2 Model Contact Problems

We consider three model problems depicted in Fig. 1. First two are scalar problems consisting of two membranes in mutual contact at adjacent edges.

© Springer International Publishing Switzerland 2016
T. Kozubek et al. (Eds.): HPCSE 2015, LNCS 9611, pp. 101–115, 2016.
DOI: 10.1007/978-3-319-40361-8_7

The solution $u(x, y)$ can be interpreted as a vertical displacement of two membranes stretched by normalized horizontal forces and pressed together by vertical forces with density $f(x, y)$. The inequality constraints result from requiring nonpenetration of the adjacent edges of the membranes, with the edge of the right membrane above the edge of the left membrane and by pressing the left membrane down by the right one at the contact points. The first problem, where the right membrane has its right edge fixed, is coercive (Fig. 1a, b). The second problem is semicoercive since the right membrane is completely floating (Fig. 1c, d).

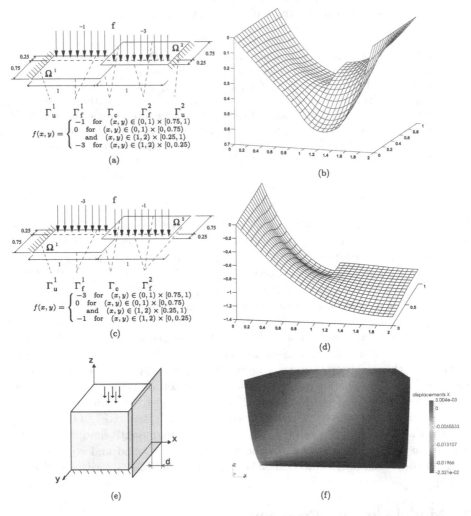

Fig. 1. Model problems: coercive (1a, b) and semicoercive (1c, d) scalar contact problem of two membranes, and elastic cube with a rigid obstacle (1e, f) – problem specification (left) and solution (resulting displacements, right) (Color figure online).

As a model 3D linear elasticity contact problem, we consider an elastic cube with the bottom face fixed, the top one loaded with a vertical surface force directed downwards, and the right one in contact with a rigid obstacle (Fig. 1e, f). The loading force density is $f_z = 465\,\mathrm{N/mm}^2$, Young's modulus $E = 2 \cdot 10^5\,\mathrm{MPa}$, Poisson's ratio $\mu = 0.33$.

3 Solution Process

This paper presents the process of solving contact problems using the PERMON Toolbox (Fig. 2). The body is decomposed in a non-overlapping way using the FETI methodology. FETI methods represent non-overlapping DDMs. They can be divided into three parts: (1) meshing part, (2) assembly part and (3) algebraic part.

Let us describe the first part. Mesh partitioning is performed first, then degrees of freedom (DOFs) on submesh interfaces are uniquely numbered (we call the resulting numbering the "undecomposed numbering"), DOFs on sub-mesh interfaces are copied to each respective submesh, and finally the DOFs are renumbered to restore global uniqueness (resulting in the "decomposed number-ing"). The resulting submeshes are completely self-contained, i.e. there is no cell overlapping or ghost-layer.

The second part, FEM assembly of algebraic objects, is performed completely separately for each submesh using any existing (even sequential) FEM code. Stiffness matrix and load vector of each submesh form a "local problem"

The third part is essentially a mathematical approach how to deal with the mesh decomposition described above so that a correct solution, continuous across submesh interfaces, is obtained. Typically, only this final part is actually called a non-overlapping DDM. To "glue" the subdomains together, the solver needs to know a mapping between the duplicated DOFs on neighbouring submesh inter-faces. This mapping results from the first part, and is simply a many-to-one mapping from the decomposed numbering to the undecomposed one provided the submesh interfaces conform. This mapping is a minimal additional informa-tion needed in contrast to "parallel linear system solvers" that just act on one distributed global stiffness matrix and load vector which come from an "unde-composed" mesh.

The first and second parts can be covered by the PermonCube and Permon-Membrane packages (Sect. 4), while the third one is realized using the Permon-FLLOP package (Sect. 5).

4 PermonMembrane and PermonCube

For rapid development and testing of our solvers, PermonMembrane and Per-monCube packages [26] were developed. They implement the first and second parts in a massively parallel way for simple benchmarks generated in run-time. PermonCube is similar by focus to the software package Pamgen [19, 30]. Although it provides so far only a cubical mesh, the FEM part of the code does

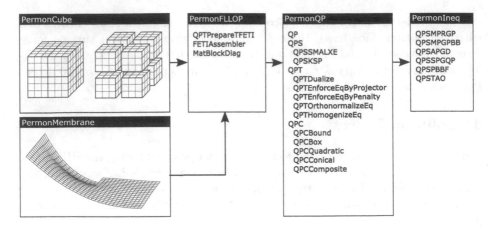

Fig. 2. PERMON life cycle.

not rely on this specific type of mesh, and works with that as if it were an unstructured mesh, simulating decomposed FEM processing of real world problems. Extending the first part to real world meshes is work in progress. Moreover, we strive to support widely available FEM libraries such as Elmer [27].

The parallel mesh generation is controlled by two groups of parameters. In PermonCube, the number of subdomains is managed by parameters X, Y, Z, and similarly the number of elements per subdomain is given by x, y, z (both considered in the respective axis directions). In PermonMembrane, the situation is similar, only parameters Z and z are missing. The decomposition parameter H and the discretization parameter h is given as $H = \frac{L}{X}$ and $h = \frac{H}{x}$, respectively, where L denotes the size of the whole domain.

Essential data, generated by PermonCube, PermonMembrane or any other FEM software, are the subdomain stiffness matrices $\mathbf{K}^s$ and the subdomain right-hand side vectors $\mathbf{f}^s$, $s = 1,\dots,N_S$ where N_S denotes the total number of subdomains, $N_S = XYZ$ for PermonCube and $N_S = 2XY$ for PermonMembrane. In the DDM context, an additional object, the previously described local-to-global interface DOF mapping $l2g$, has to be created. These data are passed to PermonFLLOP, described in the next section.

5   PermonFLLOP

Our PermonFLLOP package implements the algebraic part of non-overlapping DDMs of the FETI type. We shall firstly briefly introduce the FETI-1 and Total-FETI (TFETI) methods.

FETI-1 [12–14,23] is a non-overlapping DDM [16]. Thus, it is based on decomposing the original spatial domain into non-overlapping subdomains. They are "glued together" by Lagrange multipliers which have to satisfy certain equality constraints, discussed later. The original FETI-1 method assumes that the boundary subdomains inherit Dirichlet conditions from the original problem where the conditions are embedded into the linear system arising from FEM. This means physically that subdomains, whose interfaces intersect the Dirichlet boundary, are fixed while others are kept floating; in the linear algebra speech, the corresponding subdomain stiffness matrices are non-singular and singular, respectively.

The basic idea of the TFETI method [5,7,32] is to keep all the subdomains floating and enforce the Dirichlet boundary conditions by means of a constraint matrix and Lagrange multipliers, similarly to the gluing conditions along subdomain interfaces. This simplifies the implementation of the stiffness matrix pseudoinverse. The key point is that kernels of subdomain stiffness matrices are known a priori, have the same dimension and can be formed without any computation from the mesh data. Furthermore, each local stiffness matrix can be regularized cheaply, and the inverse of the resulting nonsingular matrix is a pseudoinverse of the original singular one [4,6].

Let us consider a partitioning of the global domain Ω into N_S subdomains $\Omega^s, s = 1, \ldots, N_S$. To each subdomain Ω^s there corresponds the subdomain stiffness matrix $\mathbf{K}^s$, the subdomain nodal load vector $\mathbf{f}^s$, the matrix $\mathbf{R}^s$ whose columns span the nullspace (kernel) of $\mathbf{K}^s$, and the constraint matrix $\mathbf{B}^s$. The latter consists of the equality and inequality parts, $(\mathbf{B}^s)^T = [(\mathbf{B}_E^s)^T \ (\mathbf{B}_I^s)^T]$. The equality part is $(\mathbf{B}_E^s)^T = [(\mathbf{B}_g^s)^T \ (\mathbf{B}_d^s)^T]$, where $\mathbf{B}_g^s$ is a signed Boolean matrix defining connectivity of the subdomain Ω^s with all its neighbouring subdomains, and $\mathbf{B}_d$ is a Boolean matrix describing Dirichlet boundary conditions (empty if the TFETI approach is not used). The inequality part $\mathbf{B}_I^s$ (possibly empty) describes linearized non-penetration conditions [7] of the subdomain Ω^s on the corresponding part of the contact zone. The global constraint right-hand side vector $\mathbf{c}$ and vector of Lagrange multipliers $\boldsymbol{\lambda}$ possess analogous structure.

The local objects $\mathbf{K}^s, \mathbf{f}^s, \mathbf{R}^s$ and $\mathbf{B}^s$ constitute global objects

$$\mathbf{K} = \mathrm{diag}(\mathbf{K}^1, \ldots, \mathbf{K}^{N_S}), \quad \mathbf{B}_E = [\mathbf{B}_E^1, \ldots, \mathbf{B}_E^{N_S}], \quad \mathbf{B} = [\mathbf{B}^1, \ldots, \mathbf{B}^{N_S}] = \begin{bmatrix} \mathbf{B}_E \\ \mathbf{B}_I \end{bmatrix},$$

$$\mathbf{R} = \mathrm{diag}(\mathbf{R}^1, \ldots, \mathbf{R}^{N_S}), \quad \mathbf{B}_I = [\mathbf{B}_I^1, \ldots, \mathbf{B}_I^{N_S}], \quad \mathbf{f} = [(\mathbf{f}^1)^T, \ldots, (\mathbf{f}^{N_S})^T]^T,$$

where diag means a block-diagonal matrix consisting of the diagonal blocks between parentheses. Note that columns of $\mathbf{R}$ also span the kernel of $\mathbf{K}$. The global discrete form of the contact problem can be written as the primal QP

$$\min \frac{1}{2}\mathbf{u}^T\mathbf{K}\mathbf{u} - \mathbf{u}^T\mathbf{f} \quad \text{s.t.} \quad \mathbf{B}_E\mathbf{u} = \mathbf{c}_E \quad \text{and} \quad \mathbf{B}_I\mathbf{u} \leq \mathbf{c}_I \tag{1}$$

with $\mathbf{c}_E$ and $\mathbf{c}_I$ denoting prescribed gaps for equality and inequality constraints. Let us apply the convex QP duality theory and establish the following notation

$$\mathbf{F} = \mathbf{B}\mathbf{K}^\dagger\mathbf{B}^T, \quad \mathbf{G} = \mathbf{R}^T\mathbf{B}^T, \quad \mathbf{d} = \mathbf{B}\mathbf{K}^\dagger\mathbf{f}, \quad \mathbf{e} = \mathbf{R}^T\mathbf{f},$$

where $\mathbf{K}^\dagger$ denotes a pseudoinverse of $\mathbf{K}$, satisfying $\mathbf{K}\mathbf{K}^\dagger\mathbf{K} = \mathbf{K}$. We obtain the dual QP

$$\min \frac{1}{2}\boldsymbol{\lambda}^T\mathbf{F}\boldsymbol{\lambda} - \boldsymbol{\lambda}^T\mathbf{d} \quad \text{s.t.} \quad \mathbf{G}\boldsymbol{\lambda} = \mathbf{e} \quad \text{and} \quad \boldsymbol{\lambda}_I \geq \mathbf{o}. \tag{2}$$

After several manipulations [7] we get the final QP, suitable for numerical solution,

$$\min \frac{1}{2}\lambda^T \mathbf{PFP}\lambda - \lambda^T \mathbf{Pd} \quad \text{s.t.} \quad \mathbf{G}\lambda = \mathbf{o} \quad \text{and} \quad \lambda_I \geq -\widetilde{\lambda}, \tag{3}$$

where

$$\mathbf{P} = \mathbf{I} - \mathbf{Q} \quad \text{and} \quad \mathbf{Q} = \mathbf{G}^T(\mathbf{GG}^T)^{-1}\mathbf{G}$$

denote the orthogonal projectors onto the kernel of $\mathbf{G}$ and image of $\mathbf{G}^T$, respectively, and $\widetilde{\lambda}$ denotes an arbitrary vector satisfying the equality constraints of (2).

Let us show how PermonFLLOP is implemented from the user's perspective. First of all, it takes from the FEM software the subdomain stiffness matrices $\mathbf{K}^s$ and the subdomain load vectors $\mathbf{f}^s$ as sequential data for each subdomain Ω^s, $s = 1, \ldots, N_S$. Note that we assume here each processor core owns only one subdomain, PermonFLLOP has nevertheless an experimental feature of allowing more than one subdomain per core, i.e. an array of $\mathbf{K}^s$ and $\mathbf{f}^s$ is passed per subdomain. PermonFLLOP enriches the independent subdomain data with the global context so that $\mathbf{K}$ and $\mathbf{f}$ are effectively created from $\mathbf{K}^s$ and $\mathbf{f}^s$, respectively.

The "gluing" signed Boolean matrix $\mathbf{B}_g$ is created based on the local-to-global mapping $l2g$ [31]. The FEM software can skip the processing of the Dirichlet conditions and rather hand it over to PermonFLLOP, resulting in greater flexibility. PermonFLLOP allows to enforce Dirichlet boundary conditions either by the constraint matrix $\mathbf{B}_d$ (TFETI approach), or by a classical technique of embedding them directly into $\mathbf{K}$ and $\mathbf{f}$ (FETI-1 approach). It is also possible to mix these two approaches.

Furthermore, PermonFLLOP assembles the nullspace matrix $\mathbf{R}$ using one of the following options. The first option is to use a numerical approach [16], and the second one is to generate $\mathbf{R}$ as rigid body modes from the mesh nodal coordinates [22]. The latter is typical for TFETI and is considered here.

Currently, PermonFLLOP requires $\mathbf{B}_I$ and $\mathbf{c}_I$ from the caller. We strive to overcome this limitation in the future so that the non-penetration conditions will be specified in a way more natural for engineers. Listing 1.1 shows how a FEM software (such as PermonCube) typically calls PermonFLLOP to solve a decomposed contact problem.

```
Mat Ks,BIs;   Vec fs,cI,coords;   IS l2g,dbcis;   MPI_Comm comm;   FLLOP fllop;

/* Generate the data. */

/* Create FLLOP living in communicator comm. */
FllopCreate(comm, &fllop);

/* Set the subdomain stiffness matrix and load vector. */
FllopSetStiffnessMatrix(fllop, Ks);
FllopSetLoadVector(fllop, fs);

/* Set the local-to-global mapping for gluing. */
FllopSetLocalToGlobalMapping(fllop, l2g);

/* Specify the Dirichlet conditions in the local numbering
   and tell FLLOP to enforce them by means of the B matrix. */
```

```
FllopAddDirichlet(fllop, dbcis, FETI_LOCAL, FETI_DBC_B);

/* Set vertex coordinates for rigid body modes. */
FllopSetCoordinates(fllop, coords);

/* Set the non-penetration inequality constraints. */
FllopSetIneq(fllop, BIs, cI);

FllopSolve(fllop);
```

Listing 1.1. PermonCube calls PermonFLLOP

In the `FllopSolve` function, PermonFLLOP passes the global primal data $\mathbf{K}$, $\mathbf{f}$, $\mathbf{B}$ and $\mathbf{R}$ to PermonQP (Sect. 6), calls a specific series of QP transforms provided by PermonQP, resulting in the bound and equality constrained QP (3), i.e. (5) with $\mathbf{A} = \mathbf{PFP}$, $\mathbf{b} = \mathbf{Pd}$, $\mathbf{C} = \mathbf{G}$, $\ell = -\tilde{\lambda}_I$, and $\mathbf{x} = \lambda$, which is then solved with the `QPSSolve` function. Listing 1.2 presents a sketch of the `FllopSolve` function.

Open source DDM codes are relatively rare. Let us mention the Multilevel BDDC solver library (BDDCML) by Šístek et al. [28,29], PETSc BDDC preconditioner implementation by S. Zampini [1], and the HPDDM code by Jolivet et al. [20,21].

```
/* FllopSolve() function */

/* Subdomain data. */
Mat Ks,BIs,Bgs,Bds,Rs; Vec fs;
/* Global data. */
Mat K, BI, Bg, Bd, R ; Vec f, cI, cd;
/* QP problem, QP solver. */
QP qp; QPS qps;

/* Create a QP data structure. */
QPCreate(comm, &qp);

/* Globalise the data. */
MatCreateBlockDiag(Ks, &K);
MatCreateBlockDiag(Rs, &R);
MatMerge(Bgs, &Bg); MatMerge(Bds, &Bd);
MatMerge(BIs, &BI); VecMerge(fs, &f);

/* Set the QP data. */
QPSetOperator(qp, K);
QPSetOperatorNullspace(qp, R);
QPSetRHS( qp, f);
QPAddEq( qp, Bg, NULL); // NULL means zero vector
QPAddEq( qp, Bd, cd);
QPSetIneq(qp, BI, cI);
```

```
/* Basic sequence of QP transforms
   giving (T)FETI method.
   QPTFetiPrepare() can be used
   instead for convenience.
   QP chain is created in backend. */
QPTScale(qp);
QPTDualize(qp);
QPTScale(qp);
QPTHomogenizeEq(qp);
QPTEnforceEqByProjector(qp);

/* Create a PermonQP solver. */
QPSCreate(comm, &qps);

/* Set the QP to be solved. */
QPSSetQP(qps, qp);

/* Solve, i.e. hand over to PermonQP.
   The last QP in the chain is solved.
*/
QPSSolve(qps);
```

Listing 1.2. PermonFLLOP calls PermonQP

6 PermonQP

PermonQP [18] allows solving QPs with an SPS Hessian and any combination of linear equality and inequality constraints including unconstrained QP. It provides a basic framework for QP solution (data structures, transformations, and supporting functions), a wrapper of PETSc KSP linear solvers for solving unconstrained and equality-constrained QP, a variant of the augmented Lagrangian method called SMALBE discussed later in Sect. 8, and several concrete solvers for bound constrained minimization (PermonIneq) – here we consider the MPRGP algorithm (Sect. 7). Its programming interface (API) is designed to be easy-to-use, and at the same time efficient and suitable for HPC. PermonQP is under preparation for publishing under the BSD 2-Clause license.

A QP transform derives a new QP from the given QP, so that a doubly linked list is generated where each node is a QP. The solution process is divided into the following sequence of actions:

1. QP problem specification;
2. a chain of QP transforms generating a chain of QP problems where the last one is passed to the solver;
3. automatic or manual choice of an appropriate QP solver;
4. the QP solver is called to solve the last QP in the chain;
5. a chain of reconstructions in the reverse order of QP transforms in order to get a solution of the original QP.

7   MPRGP

MPRGP (Modified Proportioning and Reduced Gradient Projection) [8,15] is an efficient algorithm for solution of convex QP with simple bounds

$$\min \frac{1}{2}\mathbf{x}^T \mathbf{A}\mathbf{x} - \mathbf{b}^T \mathbf{x} \quad \text{s.t.} \quad \mathbf{x}_I \geq \boldsymbol{\ell}, \tag{4}$$

where I denotes the index set corresponding to the inequality constrained entries of vector $\mathbf{x}$, and $\mathbf{x}_I$ denotes the subvector of $\mathbf{x}$ given by the index set I. This approach was introduced in [11]. The proportioning algorithm is combined with the gradient projections, a test to decide when to leave the active set, and three types of steps to generate a sequence of iterates $\mathbf{x}^k$ approximating the solution:

1. a proportioning step – removes indices from the active set,
2. a conjugate gradient (CG) step – generates the next approximation in the active set if the current approximation is proportional (i.e. meeting a special criterion related to chopped, free and reduced free gradients, see [8]),
3. an expansion step – defined by the free gradient projection with a fixed steplength $\bar{\alpha}$, expands the active set.

Instead of verifying the Karush-Kuhn-Tucker optimality conditions directly, the algorithm evaluates the *projected gradient* $\mathbf{g}^P$, given componentwise by

$$\mathbf{g}_i^P = \begin{cases} \mathbf{g}_i & \text{for} \quad x_i > l_i \quad \text{or} \quad i \notin I, \\ \min(\mathbf{g}_i, 0) & \text{for} \quad x_i = l_i \quad \text{and} \quad i \in I, \end{cases}$$

where x_i and l_i is the i-th component of $\mathbf{x}$ and $\boldsymbol{\ell}$, respectively, and $\mathbf{g} = \mathbf{A}\mathbf{x} - \mathbf{b}$ is the gradient of the objective function. The algorithm stops, when $\|\mathbf{g}^P\|$ is sufficiently small. MPRGP has a known rate of convergence given in terms of the spectral condition number of the Hessian, and may be comparable to the cost of solution of the corresponding unconstrained QP [8].

8   SMALBE

SMALBE (Semi-Monotonic Augmented Lagrangian with Bound and Equality) [8] is a variant of the inexact augmented Lagrangian algorithm, and can be viewed as an extension of the augmented Lagrangian method. It can be used to solve a box and equality constrained QP

$$\min \frac{1}{2}\mathbf{x}^T \mathbf{A}\mathbf{x} - \mathbf{b}^T \mathbf{x} \quad \text{s.t.} \quad \mathbf{x}_I \geq \boldsymbol{\ell} \quad \text{and} \quad \mathbf{C}\mathbf{x} = \mathbf{o}. \tag{5}$$

Particularly, such QPs arise from applying the FETI methodology to contact problems. The SMALBE algorithm is based on the outer loop refining the Lagrange multipliers $\boldsymbol{\mu}$ related to the equality constraints. In each outer iteration, the inner loop solving an auxiliary minimization problem

$$\min_{\mathbf{x}} L(\mathbf{x}, \boldsymbol{\mu}, \rho) \quad \text{s.t.} \quad \mathbf{x}_I \geq \boldsymbol{\ell} \tag{6}$$

is performed, where L is the *augmented Lagrangian* defined as

$$L(\mathbf{x}, \boldsymbol{\mu}, \rho) = \frac{1}{2}\mathbf{x}^T \mathbf{A}\mathbf{x} - \mathbf{b}^T \mathbf{x} + \boldsymbol{\mu}^T \mathbf{C}\mathbf{x} + \frac{\rho}{2}||\mathbf{C}\mathbf{x}||^2.$$

Using just a different bracketing, this inner problem is a QP with the penalized Hessian and updated right-hand side

$$\min_{\mathbf{x}} \frac{1}{2}\mathbf{x}^T (\mathbf{A} + \rho \mathbf{C}^T \mathbf{C})\mathbf{x} - (\mathbf{b} - \mathbf{C}^T \boldsymbol{\mu})^T \mathbf{x} \quad \text{s.t.} \quad \mathbf{x}_I \geq \boldsymbol{\ell}, \tag{7}$$

solvable by any solver for bound constrained QP such as MPRGP (Sect. 7). Here $\boldsymbol{\mu}$ is an approximation of the equality constraint Lagrange multipliers. The inner loop is stopped if $||\mathbf{g}^P|| < \varepsilon$ and $||\mathbf{C}\mathbf{x}|| < \varepsilon$ (the outer QP is already solved), or as early as when $||\mathbf{g}^P|| < \min\{M||\mathbf{C}\mathbf{x}||, \eta\}$, where $M > 0$ and $\eta > 0$ are algorithmic parameters.

The outer also updates the M parameter (SMALBE-M) or the penalty ρ (SMALBE-ρ) based on the increase of the augmented Lagrangian L with respect to the equality constraint Lagrange multiplier $\boldsymbol{\mu}$. SMALBE-M is preferred as it uses a fixed ρ and hence the Hessian and its spectrum is not altered. M is divided by the update constant $\beta > 1$ if the increase of the augmented Lagrangian with respect to $\boldsymbol{\mu}$ is not sufficient. Compared to the basic penalty method, the algorithm is able to find an approximation of $\boldsymbol{\mu}$ meeting the given precision with no need for a large penalty, avoiding ill-conditioning. Compared to the basic augmented Lagrangian method, the introduced adaptivity weakens the effect of the proper selection of the penalty sequence and eliminates necessity of exact solution of the inner problems. Optimality results for SMALBE were presented in [7–10].

9   Numerical Experiments

We illustrate the numerical scalability of TFETI for contact problems combined with SMALBE and MPRGP and weak parallel scalability of their PERMON

Table 1. Dimension settings for coercive and semicoercive problem with $h = H/128$.

$1/H$	# Subdomains	# DOFs	Dual dim – coercive	Dual dim – semicoercive	Kernel dim
8	128	2,130,048	32,160	31,142	128
11	242	4,027,122	61,377	59,978	242
16	512	8,520,192	130,872	128,838	512
24	1,152	19,170,432	296,144	293,094	1,152
32	2,048	34,080,768	527,976	523,910	2,048

Table 2. Dimension settings for cube in 3D with $h = H/24$.

$1/H$	# Subdomains	# DOFs	Dual dim	Kernel dim
5	125	3,472,875	469,392	750
6	216	6,001,128	832,194	1,296
8	512	14,224,896	2,035,950	3,072
10	1,000	27,783,000	4,051,602	6,000
13	2,197	61,039,251	9,055,080	13,182

implementations on the three model problems introduced in Sect. 2. The descriptions of the problems and their respective solutions have been shown in Fig. 1.

Regarding the first two problems, the coercive and semicoercive membrane problems, each of two membranes was first partitioned into subdomains with the sidelengths $H \in \{1/8, 1/11, 1/16, 1/24, 1/32\}$. The square subdomains were then discretized by the regular grids with the discretization parameter $h = H/128$, so that the ratio H/h was kept constant. The third problem, the elastic cube, was decomposed into subdomains with sidelengths $H \in \{1/5, 1/6, 1/8, 1/10, 1/13\}$ and discretized with $h = 1/20$ and again with constant H/h. In all cases, the splitting was chosen in order to get the numbers of subdomains near the powers of two. The corresponding total numbers of subdomains as well as the primal, dual and kernel dimensions can be found in Tables 1 and 2. Let us remind that the dual dimension is the total number of gluing, Dirichlet and non-penetration interface constraints, i.e. number of rows of the matrix $\mathbf{B}$.

We used the following parameters setting for the SMALBE and MPRGP algorithms:

$$M_0 = 100\|\mathbf{PFP}\|, \quad \rho = 2\|\mathbf{PFP}\|, \quad \eta = 0.1\|\mathbf{Pd}\| \quad \text{and} \quad \beta = 10,$$

where the matrix norms were approximated using the power method. These values are default in PermonQP and they have been chosen based on many comparative numerical tests. Needless to say, the optimal values for particular problems may slightly differ.

The stopping criterion was

$$\|\mathbf{g}^P\| \leq \varepsilon \|\mathbf{Pd}\| \ \wedge \ \|\mathbf{G}\boldsymbol{\lambda}\| \leq \varepsilon \|\mathbf{Pd}\|, \qquad \varepsilon = 10^{-4}.$$

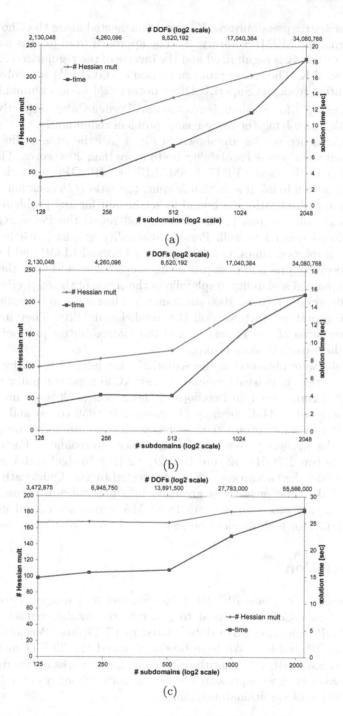

Fig. 3. Graphs of numerical and weak parallel scalability for the coercive (3a) and semicoercive (3b) membrane problems, and the cube problem (3c).

The stiffness matrix pseudoinverse $\mathbf{K}^\dagger$ was implemented using the Cholesky factorization from the MUMPS library [2]. The approach from [6] was used where the original matrix $\mathbf{K}$ is regularized and the inverse of the regularized matrix is a pseudoinverse of $\mathbf{K}$. The coarse problem (action of $(\mathbf{G}\mathbf{G}^T)^{-1}$) was solved by the LU factorization from the SuperLU_DIST library [25] in subcommunicators of size $N_S^{1/2}$ and $N_S^{2/3}$ for PermonMembrane and PermonCube, respectively, each subcommunicator solving the same coarse problem redundantly.

The performance results are shown in Fig. 3. All the graphs illustrate the numerical and weak parallel scalability up to more than 2000 cores. The numerical scalability of the used TFETI + SMALBE + MPRGP approach has been theoretically proven in [8]. It says that keeping the ratio H/h constant, the number of Hessian multiplications is bound by a constant for any problem size. The numerical scalability graphs (with circle marks) reveal the PermonQP implementation fulfils this fairly well. Parallel scalability graphs (with box marks) show the total solution times, i.e. time spent in PermonFLLOP and PermonQP including necessary pre- and post-processing steps before and after the iterative phase. Each parallel scalability graph follows the shape of the respective numerical scalability graph up to ca. 1000 subdomains. Then some worse scalable parts of the implementation start to spoil the parallel scalability. They include e.g. the implementation of the $\mathbf{B}$ actions and the matrix-matrix product $\mathbf{G} * \mathbf{G}^T$. Improving these parts is work-in-progress.

The results were obtained at the ARCHER, the latest UK National Supercomputing Service. It is based around a Cray XC30 supercomputer with 4920 nodes, 118,080 cores and 1.56 Petaflops of theoretical peak performance (4544 standard nodes with 64 GB memory (12 groups, 109,056 cores) and 376 nodes with 128 GB memory (1 group, 9,024 cores)). All compute nodes are connected together in the Dragonfly topology by the Aries interconnect. Each compute node contains two 2.7 GHz, 12-core E5-2697 v2 (Ivy Bridge) series processors. Within the node, the two processors are connected by two QuickPath Interconnect (QPI) links. The memory is arranged in a non-uniform access (NUMA) form: each 12-core processor is a single NUMA region with local memory of 32 GB (or 64 GB for high-memory nodes).

10   Conclusion

We have presented our new PERMON toolbox and its packages. PermonMembrane and PermonCube were used to generate the model contact problems, PermonFLLOP generated extra data related to FETI, and PermonQP solved the resulting QP problem. We have briefly reviewed the TFETI method, and MPRGP and SMALBE QP algorithms. Finally, benchmarks of two membranes and of the elastic cube were presented to demonstrate efficiency of the PERMON tools for solution of variational inequalities.

Acknowledgements. We would like to thank our colleague Alexandros Markopoulos for developing most of the PermonCube package. This work made use of the facilities of ARCHER, the UK's national high-performance computing service, provided by The Engineering and Physical Sciences Research Council (EPSRC), The Natural Environment Research Council (NERC), EPCC, Cray Inc. and The University of Edinburgh. We also acknowledge use of Anselm and Salomon clusters, operated by IT4Innovations National Supercomputing Center, VSB - Technical University of Ostrava, Czech Republic, for development and testing of the PERMON software. This work was supported by The Ministry of Education, Youth and Sports from the National Programme of Sustainability (NPU II) project "IT4Innovations excellence in science" (LQ1602), and from the "Large Infrastructures for Research, Experimental Development and Innovations" project "IT4Innovations National Supercomputing Center" (LM2015070); by the EXA2CT project funded from the EU's Seventh Framework Programme (FP7/2007-2013) under grant agreement No. 610741; by the internal student grant competition project "PERMON toolbox development II" (SP2016/178); by the POSTDOCI II project (CZ.1.07/2.3.00/30.0055) within Operational Programme Education for Competitiveness; and by the Grant Agency of the Czech Republic (GACR) project No. 15-18274S. This work was also supported by the READEX project – the European Union's Horizon 2020 research and innovation programme under grant agreement No. 671657.

References

1. PETSc PCBDDC manual page. http://www.mcs.anl.gov/petsc/petsc-current/docs/manualpages/PC/PCBDDC.html
2. Amestoy, P., et al.: MUMPS web pages (2015). http://mumps.enseeiht.fr/index.php?page=home
3. Balay, S., Abhyankar, S., Adams, M.F., Brown, J., Brune, P., Buschelman, K., Eijkhout, V., Gropp, W.D., Kaushik, D., Knepley, M.G., McInnes, L.C., Rupp, K., Smith, B.F., Zhang, H.: PETSc web pages (2015). http://www.mcs.anl.gov/petsc
4. Brzobohatý, T., Dostál, Z., Kozubek, T., Kovář, P., Markopoulos, A.: Cholesky decomposition with fixing nodes to stable computation of a generalized inverse of the stiffness matrix of a floating structure. Int. J. Numer. Methods Eng. **88**(5), 493–509 (2011)
5. Dostál, Z., Horák, D., Kučera, R.: Total FETI - an easier implementable variant of the FETI method for numerical solution of elliptic PDE. Commun. Numer. Methods Eng. **22**(12), 1155–1162 (2006)
6. Dostál, Z., Kozubek, T., Markopoulos, A., Menšík, M.: Cholesky decomposition of a positive semidefinite matrix with known kernel. Appl. Math. Comput. **217**(13), 6067–6077 (2011)
7. Dostál, Z., Kozubek, T., Vondrák, V., Brzobohatý, T., Markopoulos, A.: Scalable TFETI algorithm for the solution of multibody contact problems of elasticity. Int. J. Numer. Methods Eng. **82**(11), 1384–1405 (2010)
8. Dostál, Z.: Optimal Quadratic Programming Algorithms, with Applications to Variational Inequalities. SOIA, vol. 23. Springer, New York (2009)
9. Dostál, Z., Horák, D.: Theoretically supported scalable FETI for numerical solution of variational inequalities. SIAM J. Numer. Anal. **45**(2), 500–513 (2007)

10. Dostál, Z., Horák, D., Kučera, R., Vondrák, V., Haslinger, J., Dobiáš, J., Pták, S.: FETI based algorithms for contact problems: scalability, large displacements and 3D coulomb friction. Comput. Methods Appl. Mech. Eng. **194**(2–5), 395–409 (2005)
11. Dostál, Z., Schöberl, J.: Minimizing quadratic functions subject to bound constraints. Comput. Optim. Appl. **30**(1), 23–43 (2005)
12. Farhat, C., Mandel, J., Roux, F.X.: Optimal convergence properties of the FETI domain decomposition method. Comput. Methods Appl. Mech. Eng. **115**, 365–385 (1994)
13. Farhat, C., Roux, F.X.: A method of finite element tearing and interconnecting and its parallel solution algorithm. Int. J. Numer. Methods Eng. **32**(6), 1205–1227 (1991)
14. Farhat, C., Roux, F.X.: An unconventional domain decomposition method for an efficient parallel solution of large-scale finite element systems. SIAM J. Sci. Stat. Comput. **13**(1), 379–396 (1992)
15. Friedlander, A., Martínez, J.M., Raydan, M.: New method for large-scale box constrained convex quadratic minimization problems. Optim. Methods Softw. **5**(1), 57–74 (1995)
16. Gosselet, P., Rey, C.: Non-overlapping domain decomposition methods in structural mechanics. Arch. Comput. Methods Eng. **13**(4), 515–572 (2006)
17. Hapla, V., et al.: PERMON (Parallel, Efficient, Robust, Modular, Object-oriented, Numerical) web pages (2015). http://industry.it4i.cz/en/products/permon/
18. Hapla, V., et al.: PermonQP web pages (2015). http://industry.it4i.cz/en/products/permon/qp/
19. Hensinger, D.M., Drake, R.R., Foucar, J.G., Gardiner, T.A.: Pamgen, a library for parallel generation of simple finite element meshes. Technical report SAND2008-1933, Sandia National Laboratories Technical Report (2008)
20. Jolivet, P., Hecht, F., Nataf, F., Prud'homme, C.: Scalable domain decomposition preconditioners for heterogeneous elliptic problems. In: Proceedings of the International Conference on High Performance Computing, Networking, Storage and Analysis, SC 2013, pp. 80:1–80:11. ACM, New York, NY, USA (2013)
21. Jolivet, P., et al.: HPDDM high-performance unified framework for domain decomposition methods. https://github.com/hpddm/hpddm
22. Kozubek, T., Vondrák, V., Menšík, M., Horák, D., Dostál, Z., Hapla, V., Kabelíková, P., Čermák, M.: Total FETI domain decomposition method and its massively parallel implementation. Adv. Eng. Softw. **60–61**, 14–22 (2013)
23. Kruis, J.: Domain Decomposition Methods for Distributed Computing. Saxe-Coburg Publications, Stirling (2006)
24. Kruis, J.: The FETI method and its applications: a review. In: Topping, B., Iványi, P. (eds.) Parallel, Distributed and Grid Computing for Engineering, vol. 21, pp. 199–216. Saxe-Coburg Publications, Stirling (2009)
25. Li, X.S., et al.: SuperLU. http://acts.nersc.gov/superlu/
26. Markopoulos, A., Hapla, V., Cermak, M., Fusek, M.: Massively parallel solution of elastoplasticity problems with tens of millions of unknowns using PermonCube and FLLOP packages. Appl. Math. Comput. **267**, 698–710 (2015)
27. Raback, P., et al.: Elmer web pages (2015). http://www.csc.fi/english/pages/elmer/
28. Šístek, J., et al.: The Multilevel BDDC solver library (BDDCML). http://users.math.cas.cz/~sistek/software/bddcml.html
29. Sousedík, B., Šístek, J., Mandel, J.: Adaptive-multilevel BDDC and its parallel implementation. Computing **95**(12), 1087–1119 (2013)

30. The Trilinos Project: PAMGEN web pages (2015). http://trilinos.org/packages/pamgen/
31. Vašatová, A., Čermák, M., Hapla, V.: Parallel implementation of the FETI DDM constraint matrix on top of PETSc for the PermonFLLOP package. In: Wyrzykowski, R., Deelman, E., Dongarra, J., Karczewski, K., Kitowski, J., Wiatr, K. (eds.) Parallel Processing and Applied Mathematics. LNCS, vol. 9573, pp. 150–159. Springer, Heidelberg (2015)
32. Čermák, M., Hapla, V., Horák, D., Merta, M., Markopoulos, A.: Total-FETI domain decomposition method for solution of elasto-plastic problems. Adv. Eng. Softw. 84, 48–54 (2015)

Many Core Acceleration
of the Boundary Element Method

Michal Merta[1], Jan Zapletal[1](✉), and Jiri Jaros[2]

[1] IT4Innovations National Supercomputing Center, VŠB-Technical University
of Ostrava, 17. listopadu 15/2172, 708 33 Ostrava, Czech Republic
{michal.merta,jan.zapletal}@vsb.cz
[2] Department of Computer Systems, Brno University of Technology,
Božetěchova 1/2, 612 66 Brno, Czech Republic
jarosjir@fit.vutbr.cz

Abstract. The paper presents the boundary element method accelerated by the Intel Xeon Phi coprocessors. An overview of the boundary element method for the 3D Laplace equation is given followed by the discretization and its parallelization using OpenMP and the offload features of the Xeon Phi coprocessor are discussed. The results of numerical experiments for both single- and double-layer boundary integral operators are presented. In most cases the accelerated code significantly outperforms the original code running solely on Intel Xeon processors.

Keywords: Boundary element method · Intel Many Integrated Core architecture · Acceleration · OpenMP parallelization

1 Introduction

The necessity to prepare existing scientific codes for the upcoming many-core era has been discussed in numerous works since the introduction of Intel's Many Integrated Core (MIC) architecture in 2010 [3,4]. According to the June edition of 2015 Top 500 list, 35 out of 500 most powerful supercomputers are equipped with the first generation of Intel Xeon Phi coprocessor (Knights Corner, KNC). This number is expected to grow with the introduction of the second (Knights Landing, KNL) and third (Knights Hill, KNH) generations. Announced systems utilizing these architectures include the NERCS's next supercomputing system Cori with the theoretical peak performance of 30 PFLOPS or Argonne's future Aurora supercomputer with the estimated peak performance of up to 450 PFLOPS, which are expected to be operational as of 2016 and 2018, respectively.

The current generation of the Xeon Phi coprocessors features up to 61 cores with four hyper-threads per core and 350 GB/s memory bandwidth. The extended 512-bit wide AVX vector unit provides support for a concurrent SIMD operation on eight double-precision or 16 single-precision operands. The coprocessor offers two main usage modes: the native execution model, when the application directly runs on the coprocessor, and the offload model, when the application runs on the host

© Springer International Publishing Switzerland 2016
T. Kozubek et al. (Eds.): HPCSE 2015, LNCS 9611, pp. 116–125, 2016.
DOI: 10.1007/978-3-319-40361-8_8

processor which offloads certain parts of the code to the coprocessor (see Fig. 1). In this paper we focus on the latter approach. The future KNL and KNH architectures will not only be available in the form of coprocessor cards, but also as stand-alone host processors, which will eradicate the need for data movement via the PCIe interface.

In this paper we demonstrate the performance gain achievable by Intel Xeon Phi acceleration in the field of boundary integral equations. The boundary element method (BEM) is suitable for the solution of partial differential equations which can be formulated in the form of boundary integral equations, i.e., for which the so-called fundamental solution is known. Since BEM reduces the problem to the boundary of the computational domain, it is well suited for problems such as exterior sound scattering or shape optimization. However, the classical method has quadratic computational and memory complexity with respect to the number of surface degrees of freedom and produces fully populated system matrices. In addition, a special quadrature method has to be applied due to the singularities in the kernel of the underlying boundary integrals [12,14]. Although one can employ several fast BEM techniques to reduce the complexity to almost linear (see, e.g., [1,11–13]) an efficient implementation and parallelization of the method is necessary to enable solution of large scale problems. In [9] we have presented an explicit vectorization of BEM, whereas in [7] a new approach based on the graph theory for the system matrix distribution among MPI processes was introduced. In this paper we focus on the acceleration of the classical BEM by the Intel Xeon Phi coprocessors. Similar topic has been discussed, e.g., in [6], where the coprocessors were employed to accelerate the evaluation of the representation formula. We, on the other hand, investigate the assembly of the BEM system matrices necessary to compute the complete Cauchy data of the solution.

The structure of the paper is as follows. In Sect. 2 we introduce the model problem represented by the potential equation in 3D and present its boundary formulation. Section 3 concentrates on the discretization of the boundary integral equations introduced in Sect. 2 and discusses the parallelization approach used in the numerical experiments presented in Sect. 4. We summarize the results and conclude in Sect. 5.

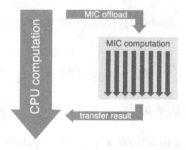

Fig. 1. Using the offload mode of Intel Xeon Phi to accelerate computation.

2 Boundary Integral Equations for the Laplace Equation

Despite its simplicity, the Laplace (or the potential) equation finds its application in numerous fields. Harmonic functions, i.e., the solutions to the Laplace equation, can be used to model the steady-state heat flow or as potentials of electrostatic or gravitational fields.

In the following we consider the mixed boundary value problem for the Laplace equation

$$\begin{cases} -\Delta u = 0 & \text{in } \Omega, \\ u = g & \text{on } \Gamma_D, \\ \dfrac{\partial u}{\partial n} = h & \text{on } \Gamma_N. \end{cases} \tag{1}$$

In (1), $\Omega \subset \mathbb{R}^3$ denotes a bounded Lipschitz domain with the boundary composed of two components $\partial\Omega = \overline{\Gamma_D} \cup \overline{\Gamma_N}$, $\Gamma_D \cap \Gamma_N = \emptyset$, and the boundary data $g \in H^{1/2}(\Gamma_D)$, $h \in H^{-1/2}(\Gamma_N)$. The solution to the Laplace equation is given by the representation formula [8,12,14]

$$\begin{aligned} u(\boldsymbol{x}) = & \int_{\Gamma_D} \gamma^1 u(\boldsymbol{y})\, v(\boldsymbol{x},\boldsymbol{y})\,\mathrm{d}s_{\boldsymbol{y}} + \int_{\Gamma_N} h(\boldsymbol{y})\, v(\boldsymbol{x},\boldsymbol{y})\,\mathrm{d}s_{\boldsymbol{y}} \\ & - \int_{\Gamma_D} g(\boldsymbol{y})\, \frac{\partial v}{\partial n_{\boldsymbol{y}}}(\boldsymbol{x},\boldsymbol{y})\,\mathrm{d}s_{\boldsymbol{y}} - \int_{\Gamma_N} \gamma^0 u(\boldsymbol{y})\, \frac{\partial v}{\partial n_{\boldsymbol{y}}}(\boldsymbol{x},\boldsymbol{y})\,\mathrm{d}s_{\boldsymbol{y}} \quad \text{for } \boldsymbol{x} \in \Omega \end{aligned}$$

with the Dirichlet and Neumann trace operators γ^0 and γ^1, respectively, and the fundamental solution $v \colon \mathbb{R}^3 \times \mathbb{R}^3 \to \mathbb{R}$,

$$v(\boldsymbol{x},\boldsymbol{y}) := \frac{1}{4\pi} \frac{1}{\|\boldsymbol{x} - \boldsymbol{y}\|}.$$

The unknown Cauchy data $\gamma^0 u|_{\Gamma_N}$, $\gamma^1 u|_{\Gamma_D}$ can be obtained from the symmetric system of boundary integral equations [12,15,16]

$$\begin{aligned} (Vs)(\boldsymbol{x}) - (Kt)(\boldsymbol{x}) &= \frac{1}{2}\tilde{g}(\boldsymbol{x}) + (K\tilde{g})(\boldsymbol{x}) - (V\tilde{h})(\boldsymbol{x}) & \text{for } \boldsymbol{x} \in \Gamma_D, \\ (K^*s)(\boldsymbol{x}) + (Dt)(\boldsymbol{x}) &= \frac{1}{2}\tilde{h}(\boldsymbol{x}) - (K^*\tilde{h})(\boldsymbol{x}) - (D\tilde{g})(\boldsymbol{x}) & \text{for } \boldsymbol{x} \in \Gamma_N \end{aligned} \tag{2}$$

involving the single-layer, double-layer, adjoint double-layer, and hypersingular boundary integral operators

$$(Vq)(\boldsymbol{x}) := \int_{\partial\Omega} v(\boldsymbol{x},\boldsymbol{y})\, q(\boldsymbol{y})\,\mathrm{d}s_{\boldsymbol{y}}, \qquad (Kt)(\boldsymbol{x}) := \int_{\partial\Omega} \frac{\partial v}{\partial n_{\boldsymbol{y}}}(\boldsymbol{x},\boldsymbol{y})\, t(\boldsymbol{y})\,\mathrm{d}s_{\boldsymbol{y}},$$

$$(K^*q)(\boldsymbol{x}) := \int_{\partial\Omega} \frac{\partial v}{\partial n_{\boldsymbol{x}}}(\boldsymbol{x},\boldsymbol{y})\, q(\boldsymbol{y})\,\mathrm{d}s_{\boldsymbol{y}}, \quad (Dt)(\boldsymbol{x}) := -\gamma^1 \int_{\partial\Omega} \frac{\partial v}{\partial n_{\boldsymbol{y}}}(\boldsymbol{x},\boldsymbol{y})\, t(\boldsymbol{y})\,\mathrm{d}s_{\boldsymbol{y}},$$

respectively. The unknown functions s, t in (2) are defined as

$$\tilde{H}^{1/2}(\Gamma_N) \ni t := \gamma^0 u - \tilde{g}, \quad \tilde{H}^{-1/2}(\Gamma_D) \ni s := \gamma^1 u - \tilde{h},$$

with suitable extensions $\tilde{g}$, $\tilde{h}$ of the given boundary data g, h.

To compute the unknown functions s, t from (2) we employ the equivalent variational formulation

$$a(s, t, \psi, \varphi) = F(\psi, \varphi) \quad \text{for all } \psi \in \tilde{H}^{-1/2}(\Gamma_{\mathrm{D}}), \ \varphi \in \tilde{H}^{1/2}(\Gamma_{\mathrm{N}}) \tag{3}$$

with the bilinear form

$$a(s, t, \psi, \varphi) := \langle Vs, \psi \rangle_{\Gamma_{\mathrm{D}}} - \langle Kt, \psi \rangle_{\Gamma_{\mathrm{D}}} + \langle K^*s, \varphi \rangle_{\Gamma_{\mathrm{N}}} + \langle Dt, \varphi \rangle_{\Gamma_{\mathrm{N}}}$$

and the right-hand side

$$F(\psi, \varphi) := \left\langle \left(\frac{1}{2}I + K \right) \tilde{g}, \psi \right\rangle_{\Gamma_{\mathrm{D}}} - \langle V\tilde{h}, \psi \rangle_{\Gamma_{\mathrm{D}}} + \left\langle \left(\frac{1}{2}I - K^* \right) \tilde{h}, \varphi \right\rangle_{\Gamma_{\mathrm{N}}} - \langle D\tilde{g}, \varphi \rangle_{\Gamma_{\mathrm{N}}}.$$

3 Discretization and Parallelization

3.1 Discretization of the Problem

In order to solve the problem (3) numerically, we discretize the boundary $\partial\Omega$ into flat shape-regular triangles τ_i, $i \in \{1, \ldots, E\}$. The boundary energy spaces are discretized by continuous piecewise linear functions $(\varphi_j)_{j=1}^N \subset H^{1/2}(\partial\Omega)$ and piecewise constant functions $(\psi_i)_{i=1}^E \subset H^{-1/2}(\partial\Omega)$. Following [12], the discrete system of linear equations reads

$$\begin{bmatrix} \mathsf{V}_h^{\mathrm{DD}} & -\mathsf{K}_h^{\mathrm{DN}} \\ (\mathsf{K}_h^{\mathrm{DN}})^\mathsf{T} & \mathsf{D}_h^{\mathrm{NN}} \end{bmatrix} \begin{bmatrix} s \\ t \end{bmatrix} = \begin{bmatrix} -\mathsf{V}_h^{\mathrm{DN}} & \frac{1}{2}\mathsf{M}_h^{\mathrm{DD}} + \mathsf{K}_h^{\mathrm{DD}} \\ \frac{1}{2}(\mathsf{M}_h^{\mathrm{NN}})^\mathsf{T} - (\mathsf{K}_h^{\mathrm{NN}})^\mathsf{T} & -\mathsf{D}_h^{\mathrm{ND}} \end{bmatrix} \begin{bmatrix} h \\ g \end{bmatrix}$$

with $\mathsf{X}^{\bullet\bullet}$ denoting the restrictions of the matrices $\mathsf{X} \in \{\mathsf{V}_h, \mathsf{K}_h, \mathsf{D}_h, \mathsf{M}_h\}$,

$$\mathsf{V}_h[\ell, j] := \int_{\tau_\ell} \int_{\tau_j} v(\boldsymbol{x}, \boldsymbol{y}) \, \mathrm{d}s_{\boldsymbol{y}} \, \mathrm{d}s_{\boldsymbol{x}}, \qquad \mathsf{M}_h[\ell, i] := \int_{\tau_\ell} \varphi_i(\boldsymbol{x}) \, \mathrm{d}s_{\boldsymbol{x}},$$

$$\mathsf{K}_h[\ell, i] := \int_{\tau_\ell} \int_{\partial\Omega} \frac{\partial v}{\partial n_{\boldsymbol{y}}} (\boldsymbol{x}, \boldsymbol{y}) \, \varphi_i(\boldsymbol{y}) \, \mathrm{d}s_{\boldsymbol{y}} \, \mathrm{d}s_{\boldsymbol{x}}, \qquad \mathsf{D}_h := \mathsf{T}^\mathsf{T} \mathrm{diag}\{\mathsf{V}_h, \mathsf{V}_h, \mathsf{V}_h\}\mathsf{T}$$

to the respective parts of the boundary. Since M_h is a sparse matrix with no singularity in the integrand and D_h can be computed by a sparse transformation of the single-layer matrix V_h (see [2,12]), in the following we only concentrate on the efficient assembly of the matrices V_h, K_h.

To deal with the singularities in the surface integrals we use the technique proposed in [9,14]. Using a series of substitutions

$$\boldsymbol{F}_n \circ \ldots \circ \boldsymbol{F}_1 =: \boldsymbol{F} \colon [0, 1]^4 \to \tau_\ell \times \tau_j,$$

$$\boldsymbol{F}(z_1, z_2, z_3, z_4) = (\boldsymbol{x}, \boldsymbol{y}), \quad \mathrm{d}s_{\boldsymbol{y}} \, \mathrm{d}s_{\boldsymbol{x}} = \boldsymbol{S}(z_1, z_2, z_3, z_4) \, \mathrm{d}z_1 \, \mathrm{d}z_2 \, \mathrm{d}z_3 \, \mathrm{d}z_4,$$

the matrix entries for V_h (similarly for K_h) read

$$\mathsf{V}_h[\ell, j] = \int_0^1 \int_0^1 \int_0^1 \int_0^1 v(\boldsymbol{F}(z_1, z_2, z_3, z_4)) \, \boldsymbol{S}(z_1, z_2, z_3, z_4) \, \mathrm{d}z_1 \, \mathrm{d}z_2 \, \mathrm{d}z_3 \, \mathrm{d}z_4.$$

Since the transformed integrand $h := (v \circ \boldsymbol{F})\boldsymbol{S}$ is analytic, the values can be computed by a standard tensor Gauss integration scheme

$$\sum_{m=1}^{k}\sum_{n=1}^{k}\sum_{o=1}^{k}\sum_{p=1}^{k} \omega_m \, \omega_n \, \omega_o \, \omega_p \, h(z_m, z_n, z_o, z_p). \tag{4}$$

The substitution $\boldsymbol{F}$ is different for τ_ℓ, τ_j being identical, sharing exactly one edge, one vertex, or being disjoint. Note that for well separated triangles, say,

$$\|\boldsymbol{t}_\ell - \boldsymbol{t}_j\| > \eta \max\{\text{diam}\,\tau_\ell, \text{diam}\,\tau_j\}$$

with the centres of gravity $\boldsymbol{t}_\ell$, $\boldsymbol{t}_j$ and a suitable coefficient η the kernel function is smooth and the entries can be computed directly by a triangle Gauss integration scheme (see [12], Appendix C)

$$\sum_{m=1}^{k}\sum_{n=1}^{k} \omega_m \, \omega_n \, v(\boldsymbol{x}_m, \boldsymbol{y}_n). \tag{5}$$

3.2 Parallelization and Acceleration of the System Matrix Assembly

The simplified algorithm of the system matrix assembly on a CPU is depicted in Listing 1.1. The algorithm iterates through the couples of elements and assembles local system matrices. The Gaussian quadrature (5) is used for well separated elements; in other cases we employ the scheme (4).

```
1  #pragma omp parallel for
2  for (int i = 0; i < nElements; i++) {
3      for (int j = 0; j < nElements; j++) {
4          if (areElementsDistant(i, j)) {
5              getLocalMatrixFarfield(i, j, localMatrix);
6          } else {
7              getLocalMatrixNearfield(i, j, localMatrix);
8          }
9          globalMatrix.add(i, j, localMatrix);
10     }
11 }
```

Listing 1.1. Simplified CPU computation of the system matrix

After the assembly of the local matrix its contribution is added to the appropriate positions of the global system matrix (line 9 of the listing). Since the computation is done in parallel using OpenMP, the update of the global matrix has to be treated using atomic operations.

To accelerate the computation we distribute the workload among the available accelerators and the CPU by splitting the matrix into $N_{\text{MIC}} + 1$ horizontal blocks. Dimensions of the blocks should follow the ratio between the theoretical peak performance of the Xeon Phi and the host. The computation on the CPU and the coprocessors is performed simultaneously (see Fig. 1). Moreover, to keep the computation on the coprocessors as simple as possible, the accelerators only take care of the quadrature over disjoint elements. The quadrature over close and identical elements are computed on the CPU.

The process consists of several steps:

1. Preallocation of data structures (mainly the system matrix) on the host.
2. Sending the necessary data from the host to the coprocessor (mesh elements, nodes, normals, etc.).
3. Parallel computation on the coprocessor and the host.
4. Sending the partially computed system matrix from the coprocessor to the host.
5. Combination of the host and coprocessor data.

Since the amount of memory on Xeon Phi is usually lower than on the host, the matrix blocks assigned to the coprocessor are further split into chunks of smaller dimensions in order to tackle large problems. Moreover, splitting the matrix enables us to use the technique of double buffering to overlap the data transfer of a matrix chunk with the computation of the subsequent one.

The simplified source code of the offload is provided in Listing 1.2. The underlying data structures are extracted from C++ objects in order to send raw data to the coprocessor using the `#pragma offload target(mic)` statement provided by the Intel Compiler (see lines 1–11 of the listing). The parallelization on the coprocessor is performed using the ordinary OpenMP pragmas. The singular integration is skipped and left for the host. The asynchronous offloaded computation is enabled by the `signal` clause. After the data is sent to the coprocessor the CPU continues in the parallel quadrature over its portion of elements.

```
1  // get pointers to raw data
2  double * nodes  = mesh->getNodes();
3  int * elements = mesh->getElements();
4  double * matrixData = globalMatrix.getData();
5  char s;
6
7  // initialize offload region
8  #pragma offload target(mic) signal(&s) \
9      in(nodes : length(3 * nNodes)) \
10     in(elements : length(3 * nElements)) \
11     out(matrixData : length(dataLength))
12     {
13         #pragma omp parallel for
14         for (int i = myElemStart; i < myElemEnd; i++) {
15             for (int j = 0; j < nElements; j++) {
16                 if (areElementsDistant(i, j)) {
17                     getLocalMatrixFarfieldMIC(i, j, nodes,
       elements,
18                         localMatrix);
19                 } else {
20                     continue;
21                 }
22                 addToGlobalmatrix(i, j, localMatrix,
       matrixData);
23             }
24         }
25     }
26  // simultaneous computation on the host CPU
27  ...
28  // receive data from the coprocessor
29  #pragma offload target(mic:0) wait(&s)
30  // combine the data from the host and the coprocessor
```

Listing 1.2. Simplified offloaded computation of the system matrix

After finishing its work the CPU waits for the data from the coprocessor (see line 29). Finally, the contributions from the host processor and the coprocessor are combined.

Note that the method assembling the local matrix on the Intel Xeon Phi coprocessor has to be appropriately modified. For optimal memory movement the data have to be aligned to 64 byte boundaries. Moreover, the original loops over quadrature points are manually unrolled and vectorization is assisted using the simd pragma.

```
1 int out = nOutQPoints;
2 int in = nInQPoints;
3 double value = 0.0;
4 for (int i = 0; i < out; i++) {
5     for (int j = 0; j < in; j++) {
6         value += ...;
7     }
8 }
```

Listing 1.3. Gauss quadrature on CPU

```
1 int p = nOutQPoints*nInQPoints;
2 double value = 0.0;
3 #pragma simd vectorlength(8) \
4     reduction(+:value)
5 for (int i = 0; i < p; i++) {
6     value += ...;
7 }
8
```

Listing 1.4. Gauss quadrature on MIC

Although the compiler automatically vectorizes the inner loop in Listing 1.3, the longer loop in Listing 1.4 is more suitable for the extended SIMD registers on the coprocessor.

4   Numerical Experiments

The following numerical experiments were carried out on the MIC-accelerated nodes of the Salomon cluster at the IT4Innovations National Supercomputing Center, Czech Republic. The nodes are equipped with two 12-core Intel Xeon E5-2680v3 processors, 128 GB of RAM, and two Intel Xeon Phi 7120P coprocessor cards. Each coprocessor offers 61 cores running at 1.238 GHz, four hyperthreads per core, and extended 512-bit vector registers. On the coprocessor 16 GB of memory is available with the maximum memory bandwidth of 352 GB/s.

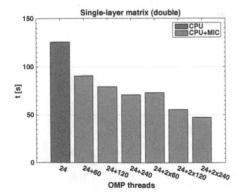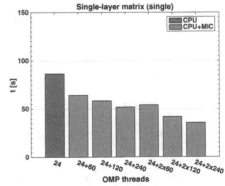

Fig. 2. Assembly of V_h accelerated by the Intel Xeon Phi coprocessors (Color figure online).

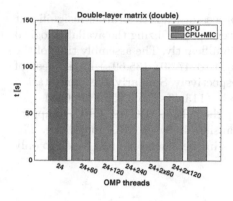 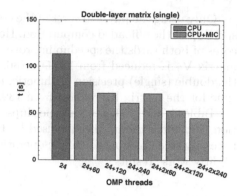

Fig. 3. Assembly of K_h accelerated by the Intel Xeon Phi coprocessors (Color figure online).

The coprocessor offers 1.2 TFLOPS of computing power, which is theoretically 1.26 times the power provided by the pair of the Intel Xeon processors (assuming 16 FMA instructions per cycle on both architectures [17]).

The experiments were performed using the Intel Compiler, version 15. The thread affinity on the coprocessor is set to `balanced`. We enable the offload runtime to allocate memory with 2 MB pages using the `MIC_USE_2MB_BUFFERS=100k` environmental variable. This significantly improves the performance since the coprocessors have to allocate data for large full system matrices. We test the assembly of the system matrices for a mesh with 81920 surface elements and 40962 nodes, thus $V_h \in \mathbb{R}^{81920 \times 81920}$ and $K_h \in \mathbb{R}^{81920 \times 40962}$.

The results of the numerical experiments are depicted in Figs. 2 and 3. We measure the assembly times on the coprocessor using the multiples of 60 threads in order to leave a core for the operating system. We compare the partially offloaded computation to the parallel assembly on 24 cores of the host's Intel Xeon processors. The workload was kept balanced among the coprocessors and CPU by adjusting the size of the matrix blocks.

Table 1. Speedup of CPU+MIC vs. CPU assembly for V_h.

# threads	24 + 60	24 + 120	24 + 240	$24 + 2 \times 60$	$24 + 2 \times 120$	$24 + 2 \times 240$
Double	1.39	1.59	1.77	1.72	2.27	2.66
Single	1.35	1.48	1.66	1.60	2.07	2.43

Table 2. Speedup of CPU+MIC vs. CPU assembly for K_h.

# threads	24 + 60	24 + 120	24 + 240	$24 + 2 \times 60$	$24 + 2 \times 120$	$24 + 2 \times 240$
Double	1.28	1.46	1.77	1.41	2.04	2.43
Single	1.37	1.59	1.87	1.60	2.16	2.54

Only using the physical cores of the coprocessor leads to a rather insignificant speedup of the offloaded computation. However, by utilizing the available logical cores of both cards the speedup improves significantly. The assembly time of the matrix V_h is reduced from 125.56 s (86.55 s) to 47.30 s (35.69 s) in the case of the double (single) precision arithmetic, respectively. Similarly, the computation time for the matrix K_h reduces from 139.83 s (113.43 s) to 57.50 s (44.64 s).

Tables 1 and 2 depict the speedups of the CPU+MIC vs. CPU computation (including the overhead caused by transfers between MIC and CPU). The necessity to employ more atomic operations in the assembly of K_h leads to only slightly worse results.

5   Conclusion

We have presented the boundary element method for the Laplace equation accelerated by the Intel Xeon Phi coprocessors. Sample codes describing the acceleration using the offload feature of the coprocessors were provided. The accelerated code featuring asynchronous CPU/coprocessor computation was tested on the Intel Xeon Phi 7120P cards in combination with two 12-core Intel Xeon processors. We have achieved a reasonable speedup in comparison with the non-accelerated code.

Although the numerical experiments were only performed for a relatively small problem, the full assembly of similar matrices can be used in boundary element tearing and interconnecting methods (BETI) [5, 10], where the computational domain is divided into many small subdomains. The global algorithm usually requires accurate solution of local problems, which can be achieved by direct solvers with full matrices.

In addition to the classical BEM and BETI we turn our attention to the acceleration of the adaptive cross approximation method (ACA), which will enable the solution of engineering problems of large dimensions. Moreover, the presented code can be relatively easily extended to support the accelerated solution of the problems of elastostatics.

Acknowledgments. This work was supported by the IT4Innovations Centre of Excellence project (CZ.1.05/1.1.00/02.0070), funded by the European Regional Development Fund and the national budget of the Czech Republic via the Research and Development for Innovations Operational Programme, as well as Czech Ministry of Education, Youth and Sports via the project Large Research, Development and Innovations Infrastructures (LM2011033). MM and JZ acknowledge the support of VŠB-TU Ostrava under the grant SGS SP2015/160. JJ was supported by the research project Architecture of parallel and embedded computer systems, Brno University of Technology, FIT-S-14-2297, 2014–2016 and by the SoMoPro II Programme co-financed by the European Union and the South-Moravian Region. This work reflects only the author's view and the European Union is not liable for any use that may be made of the information contained therein.

References

1. Bebendorf, M., Rjasanow, S.: Adaptive low-rank approximation of collocation matrices. Computing **70**(1), 1–24 (2003)
2. Dautray, R., Lions, J., Amson, J.: Mathematical Analysis and Numerical Methods for Science and Technology. Integral Equations and Numerical Methods, vol. 4. Springer, Heidelberg (1999)
3. Deslippe, J., Austin, B., Daley, C., Yang, W.S.: Lessons learned from optimizing science kernels for Intel's Knights Corner architecture. Comput. Sci. Eng. **17**(3), 30–42 (2015)
4. Dongarra, J., Gates, M., Haidar, A., Jia, Y., Kabir, K., Luszczek, P., Tomov, S.: HPC programming on intel many-integrated-core hardware with MAGMA port to Xeon Phi. Sci. Program. **2015**, 11 (2015)
5. Langer, U., Steinbach, O.: Boundary element tearing and interconnecting methods. Computing **71**(3), 205–228 (2003)
6. López-Portugués, M., López-Fernández, J., Díaz-Gracia, N., Ayestarán, R., Ranilla, J.: Aircraft noise scattering prediction using different accelerator architectures. J. Supercomputing **70**(2), 612–622 (2014). http://dx.doi.org/10.1007/s11227-014-1107-z
7. Lukáš, D., Kovář, P., Kovářová, T., Merta, M.: A parallel fast boundary element method using cyclic graph decompositions. Numer. Algorithms **70**, 807–824 (2015)
8. McLean, W.: Strongly Elliptic Systems and Boundary Integral Equations. Cambridge University Press, Cambridge (2000)
9. Merta, M., Zapletal, J.: Acceleration of boundary element method by explicit vectorization. Adv. Eng. Softw. **86**, 70–79 (2015)
10. Of, G., Steinbach, O.: The all-floating boundary element tearing and interconnecting method. J. Numer. Math. **17**(4), 277–298 (2009)
11. Of, G.: Fast multipole methods and applications. In: Schanz, M., Steinbach, O. (eds.) Boundary Element Analysis. Lecture Notes in Applied and Computational Mechanics, vol. 29, pp. 135–160. Springer, Heidelberg (2007)
12. Rjasanow, S., Steinbach, O.: The Fast Solution of Boundary Integral Equations. Springer, Heidelberg (2007)
13. Rokhlin, V.: Rapid solution of integral equations of classical potential theory. J. Comput. Phys. **60**(2), 187–207 (1985)
14. Sauter, S., Schwab, C.: Boundary Element Methods. Springer Series in Computational Mathematics. Springer, Heidelberg (2010)
15. Sirtori, S.: General stress analysis method by means of integral equations and boundary elements. Meccanica **14**(4), 210–218 (1979)
16. Steinbach, O.: Numerical Approximation Methods for Elliptic Boundary Value Problems: Finite and Boundary Elements. Texts in Applied Mathematics. Springer, Heidelberg (2008)
17. Intel Xeon Phi Coprocessor Peak Theoretical Maximums. http://www.intel.com/content/www/us/en/benchmarks/server/xeon-phi/xeon-phi-theoretical-maximums.html. Accessed 9 Oct 2015

Parallel Implementation of Collaborative Filtering Technique for Denoising of CT Images

Petr Strakos[1(✉)], Milan Jaros[1,2], Tomas Karasek[1], and Tomas Kozubek[1,2]

[1] IT4Innovations, Ostrava, Czech Republic
petr.strakos@vsb.cz
[2] Department of Applied Mathematics, VSB - Technical University of Ostrava,
Ostrava, Czech Republic

Abstract. In the paper parallelization of the collaborative filtering technique for image denoising is presented. The filter is compared with several other available methods for image denoising such as Anisotropic diffusion, Wavelet packets, Total Variation denoising, Gaussian blur, Adaptive Wiener filter and Non-Local Means filter. Application of the filter is intended for denoising of the medical CT images as a part of image pre-processing before image segmentation. The paper is evaluating the filter denoising quality and describes effective parallelization of the filtering algorithm. Results of the parallelization are presented in terms of strong and weak scalability together with algorithm speed-up compared to the typical sequential version of the algorithm.

Keywords: Parallelization · Scalability · Image filtering · CT images · Denoising

1 Introduction

Image filtering provides modification or enhancement of the image. The image denoising can be considered as one of the most important techniques of image enhancement. The image noise is usually considered as something unwanted and negative to the extent that we would like to eliminate it or at least attenuate as much as possible. This is especially the case of medical images acquired by Computed Tomography (CT) [1,2] or Magnetic Resonance Imaging (MRI).

Technology of CT works with X-rays that scan the object such as human body. CT can recreate the image of the internal human organs in sequential slices. Although the technology improved a lot during the past years, significant noise artefacts are still present in the images. This is unconditionally connected with the CT scanning process itself. Technically the noise can be attenuated by the higher amount of photons emitted and subsequently acquired during the scanning process. However, the higher doses of X-rays can be lethal for the patient and thus the level of noise in CT images can be attenuated only to a certain level.

© Springer International Publishing Switzerland 2016
T. Kozubek et al. (Eds.): HPCSE 2015, LNCS 9611, pp. 126–140, 2016.
DOI: 10.1007/978-3-319-40361-8_9

During the development of Computed Tomography new directions in diagnostic medicine have appeared [1,2]. CT images could be used to validate therapeutical effectiveness in cancer diseases or help in planning of surgeries. For this purpose, 3D models of the human organs can be very useful. Creation of the 3D model from CT images requires image segmentation, which is very sensitive to the presence of noise.

Most of the denoising methods try to attenuate the noise based on some assumptions about the noise characteristics. Usually, and this holds for CT image noise also, additive white Gaussian noise is considered [3,4]. Effectivity of denoising methods is very often proportional to computational extent. Denoising of the image can be time demanding process especially in the areas of medical imaging. During the process of creation of 3D model of organs hundreds or even thousands of images have to be processed before segmentation and 3D organ reconstruction can be started. In such cases High Performance Computing (HPC) can bring significant speed-up into the whole process. To use the advantage of HPC scalable versions of the algorithms incorporated in the process have to be used.

In this paper parallelized version of the collaborative filtering method for image denoising based on [5] is presented. The described method is parallelized by dividing the flow of the algorithm into separable steps which can then run in parallel. The proposed parallelization concept is proved by significant computational speed-up.

The paper is organized in the following manner. In the second section filtering method is described. In the third section denoising capabilities of the selected method are evaluated and compared with other available methods such as Anisotropic diffusion, Wavelet packets, Total Variation denoising, Gaussian blur, Adaptive Wiener filter and Non-Local Means filter. Comparison is made on medical CT images. In the fourth section parallel implementation of the method is presented. In the fifth section results of runtime tests that compare the sequential and the parallel implementation of the algorithm are presented. The sixth and last section brings the conclusions.

2 Collaborative Filtering Method

The digital image can be considered as a sum of an original noiseless image and some additive noise

$$z(x) = y(x) + \eta(x). \tag{1}$$

In (1) x stands for the pixel coordinate that belongs to the image domain X, $x \in X$. The image domain is defined by the size of the image $n \times m$. By $y(x)$ the pixel value at coordinate x in the original noiseless image is specified and $\eta(x)$ stands for the noise added to the image at the coordinate x. As a noise zero-mean Gaussian noise with standard deviation σ is considered.

The method presented here originates in [5]. It is visually summarized in Fig. 1 and in more detail in Fig. 2, where explanation of important terms and variables is graphically represented. The main point of the method is to collect similar data, which then increase the sparse representation in transform domain.

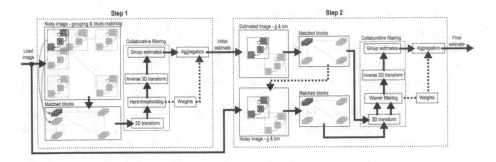

Fig. 1. Workflow of the collaborative filtering method

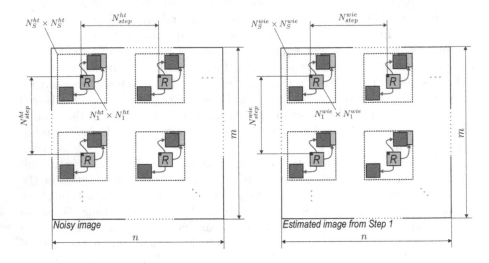

Fig. 2. Detail of the collaborative filtering method: Step 1 (left), Step 2 (right)

By data we mean data blocks or patches that are represented by image fragments within the particular image. The image fragment is denoted Z_x and specified by its location x (pixel in the left top corner). Fragments can be of general shape but here we use square fragments of size $N_1 \times N_1$. Filtering in the transform domain is then effectively achieved due to the sparsity of the data. The filtering procedure consists of two main steps. It is a creation of an initial image estimate which is characterized by hard-thresholding (superscript ht) and estimation of the final filtered image characterized by Wiener filtering (superscript wie).

2.1   Step 1 - Initial Estimate

In the following text substeps of initial estimate are described.

1. Find position x_R of each reference block by sliding with step of N_{step}^{ht} pixels in horizontal and vertical direction inside the image.

2. Calculate the group estimates. For each reference block provide the following actions.

 - Perform grouping by matching the image blocks of fixed size and square shape in selected region of size $N_S^{ht} \times N_S^{ht}$ around the reference block with the reference block. The matching is done by calculating the distance between the image blocks and the reference block as

$$d(Z_{x_R}, Z_x) = \frac{||\Upsilon'(T_{2D}^{ht}(Z_{x_R})) - \Upsilon'(T_{2D}^{ht}(Z_x))||_2^2}{(N_1^{ht})^2}. \tag{2}$$

   The reference block in location x_R is denoted as Z_{x_R}, T_{2D}^{ht} applies 2D linear transformation on the data, Υ' is the hard-thresholding operator used on the transformed data. The thresholding value for Υ' is $\lambda_{2D} \cdot \sigma$. If $d(Z_{x_R}, Z_x) \leq \tau^{ht}$, then the Z_x block matches the reference Z_{x_R}. All matched blocks together with the reference block are stacked to form the 3D array denoted by $\mathbf{Z}_{x_R}^{ht}$. The size of the array is $N_1^{ht} \times N_1^{ht} \times N_2^{ht}$, where N_2^{ht} is the maximum number of considered blocks in the stack with the smallest distance to the reference block. Limitation by N_2^{ht} helps to speed-up the calculation process.

 - Realize the collaborative filtering by 3D transform of the stacked array $\mathbf{Z}_{x_R}^{ht}$ to sparse representation (use 2D transformation on each block in the stack (horizontal direction) and 1D transformation of the transformed stack (vertical direction)) and then hard-threshold the sparse data to attenuate the noise. After the thresholding use the back-transform in reverse order to obtain denoised blocks again in spatial representation. The described procedure can be expressed as

$$\widehat{\mathbf{Y}}_{x_R}^{ht} = T_{2D}^{ht^{-1}} T_{1D}^{ht^{-1}} (\Upsilon(T_{1D}^{ht} T_{2D}^{ht}(\mathbf{Z}_{x_R}^{ht}))). \tag{3}$$

   Hard-thresholding operator Υ in (3) has threshold $\lambda_{3D} \cdot \sigma$, the array $\widehat{\mathbf{Y}}_{x_R}^{ht}$ is a group of filtered blocks $\widehat{Y}_x^{ht}$ around the particular reference block inside the image.

3. Aggregate the group estimates by weighted averaging at every pixel position in the image using the weights.

 - Initial estimate at every pixel position is computed as

$$\widehat{y}^{init}(x) = \frac{\displaystyle\sum_{x_R \in X} \sum_{x_m \in X_R} w_{x_R}^{ht} \widehat{Y}_{x_m}^{ht}(x)}{\displaystyle\sum_{x_R \in X} \sum_{x_m \in X_R} w_{x_R}^{ht} \chi_{x_m}(x)}. \tag{4}$$

   In (4) X_R is a set of block locations x_m after block matching that are spread around particular reference block located at x_R, X is the set of all possible locations of reference blocks within the image, $\widehat{Y}_{x_m}^{ht}$ is a particular image block estimated in (3), χ_{x_m} is a block support with the same size and location as $\widehat{Y}_{x_m}^{ht}$ composed of ones, $w_{x_R}^{ht}$ is the weight assigned for each group of estimates from (3). The calculation of weight $w_{x_R}^{ht}$ for each group of estimates is performed as

$$w_{x_R}^{ht} = \begin{cases} (\sigma^2 \cdot N)^{-1}, \ N \geq 1 \\ 1, \ N = 0 \end{cases}.$$ (5)

Value of N in (5) stands for the number of retained (non-zero) coefficients after the hard-thresholding performed in (3) by $\Upsilon(T_{1D}^{ht}T_{2D}^{ht}(\mathbf{Z}_{x_R}^{ht}))$.

- To reduce the border effect of each aggregated block in (4), it is beneficial to use Kaiser window with parameter β^{ht} of the same size as the aggregated block $N_1^{ht} \times N_1^{ht}$ (performed as element-by-element multiplication with $\widehat{Y}_{x_m}^{ht}$).

2.2   Step 2 - Final Estimate

To calculate final estimate following substeps are performed.

1. Find position x_R of each reference block by sliding similarly as in Step 1, but now with step N_{step}^{wie}.
2. Calculate the group estimates. For each reference block provide the following actions.
   - Perform grouping within the initial image estimate from Step 1 by matching the reference and other image blocks of size $N_1^{wie} \times N_1^{wie}$ in the area $N_S^{wie} \times N_S^{wie}$ around the reference block. The matching is done by calculating the distance between the image blocks and the reference as

$$d(\widehat{Y}_{x_R}^{init}, \widehat{Y}_x^{init}) = \frac{||\widehat{Y}_{x_R}^{init} - \widehat{Y}_x^{init}||_2^2}{(N_1^{wie})^2}.$$ (6)

The reference block in the initial image estimate at location x_R is denoted $\widehat{Y}_{x_R}$, while image block at location x is $\widehat{Y}_x$. If $d(\widehat{Y}_{x_R}^{init}, \widehat{Y}_x^{init}) \leq \tau^{wie}$, then the $\widehat{Y}_x$ block matches the reference. All matched blocks together with the reference block are stacked to 3D array $\widehat{\mathbf{Y}}_{x_R}^{init}$. The size of the array is $N_1^{wie} \times N_1^{wie} \times N_2^{wie}$, where N_2^{wie} is the maximum number of considered blocks in the stack.

- Use the same x locations that form the 3D array $\widehat{\mathbf{Y}}_{x_R}^{init}$ to form similar 3D array $\mathbf{Z}_{x_R}^{wie}$ from noisy input image.
- Compute the Wiener coefficients for every 3D group stacked above each reference block as

$$\mathbf{W}_{x_R\,ijk}^{wie} = \frac{(T_{1D}^{wie}T_{2D}^{wie}(\widehat{\mathbf{Y}}_{x_R}^{init}))_{ijk}^2}{(T_{1D}^{wie}T_{2D}^{wie}(\widehat{\mathbf{Y}}_{x_R}^{init}))_{ijk}^2 + \sigma^2}$$ (7)

Indexes i, j, k in (7) signify the operations performed over each element of the 3D array. Similarly as in Step 1, transformations T_{2D}^{wie} and T_{1D}^{wie} are used to transform $\widehat{\mathbf{Y}}_{x_R}^{init}$ into sparse representation.

- Perform collaborative Wiener filtering of $\mathbf{Z}_{x_R}^{wie}$ using element-by-element multiplication of the Wiener coefficients $\mathbf{W}_{x_R}^{wie}$ and the sparse representation of noisy data $T_{1D}^{wie}T_{2D}^{wie}(\mathbf{Z}_{x_R}^{wie})$. Combined with the inverse transformation to spatial domain the group estimates are obtained as

$$\widehat{\mathbf{Y}}_{x_R\ ijk}^{wie} = T_{2D}^{wie^{-1}}T_{1D}^{wie^{-1}}(\mathbf{W}_{x_R\ ijk}^{wie}(T_{1D}^{wie}T_{2D}^{wie}(\mathbf{Z}_{x_R}^{wie}))_{ijk}) \qquad (8)$$

The array $\widehat{\mathbf{Y}}_{x_R}^{wie}$ in (8) is composed of single block estimates $\widehat{Y}_x^{wie}$ located at x around the reference block at location x_R. To accentuate the element-by-element operations indexes i, j, k are used.

3. Aggregate the group estimates by using the weighted average of local estimates from (8).

- Final estimate at every pixel position is computed as

$$\widehat{y}^{final}(x) = \frac{\sum\limits_{x_R\in X}\sum\limits_{x_m\in X_R} w_{x_R}^{wie}\widehat{Y}_{x_m}^{wie}(x)}{\sum\limits_{x_R\in X}\sum\limits_{x_m\in X_R} w_{x_R}^{wie}\chi_{x_m}(x)}. \qquad (9)$$

Explanation of individual terms in (9) is analogous to (4). Differences are in $\widehat{Y}_{x_m}^{wie}$, which is the image block estimate resolved in (8) and in $w_{x_R}^{wie}$, that

Table 1. Parameter setting for optimal denoising capability of the collaborative filtering method; value of σ corresponds to 8-bit depth ($0 \div 255$)

		$\sigma \leq 40$	$\sigma > 40$
Step 1	T_{1D}^{ht}	1D-Haar	1D-Haar
	T_{2D}^{ht}	2D-Bior1.5	2D-DCT
	N_1^{ht}	8	12
	N_2^{ht}	16	16
	N_{step}^{ht}	3	4
	N_S^{ht}	39	39
	σ_{2D}	0	2
	σ_{3D}	2.7	2.8
	τ^{ht}	2500	5000
	β^{ht}	2.0	2.0
Step 2	T_{1D}^{wie}	1D-Haar	1D-Haar
	T_{2D}^{wie}	2D-DCT	2D-DCT
	N_1^{wie}	8	11
	N_2^{wie}	32	32
	N_{step}^{wie}	3	6
	N_S^{wie}	39	39
	τ^{wie}	400	3500
	β^{wie}	2.0	2.0

is the weight assigned for each group of estimates from (8). Calculation of weight $w_{x_R}^{wie}$ for each group of estimates is performed as

$$w_{x_R}^{wie} = \sigma^{-2}||\mathbf{W}_{x_R}^{wie}||_2^{-2} \, . \tag{10}$$

- To reduce the border effect of each aggregated block in (9) Kaiser window with parameter β^{wie} and size $N_1^{wie} \times N_1^{wie}$ is used (performed as element-by-element multiplication with $\widehat{Y}_{x_m}^{wie}$).

The described denoising method works with many parameters as shown in previous two subsections. One of the parameters that significantly influences the algorithm complexity is the total number of reference blocks, which is directly determined by the image size and the N_{step} parameter (either N_{step}^{ht} or N_{step}^{wie} based on the performed step). Optimal setting of all the parameters has been studied in [5] and is also deeply elaborated in [6]. We have adopted this setting to achieve the best denoising quality at reasonable algorithm complexity. Specific setting of the parameters is presented in Table 1.

3   Denoising Capabilities of the Collaborative Filtering Method on CT Images

As was already mentioned, since the CT images contain a lot of noise, denoising is very important technique of processing the CT images, see Fig. 3. Image segmentation [7] as a subsequent process of image treatment leading to extended possibilities in diagnostic medicine can be significantly improved by image denoising, see Fig. 3. The better denoising method is used, the more precise results in diagnostic procedures can be expected.

The selected method of collaborative filtering shows very good results in terms of denoising quality. A study showing its superior denoising potential in comparison with 13 other advanced methods is shown in [8]. Only two other algorithms which are direct modifications of the selected method gave better results in [8], but the quality improvements are only minor in comparison to the

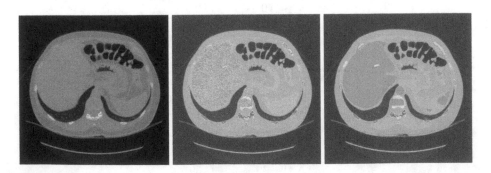

Fig. 3. From left to right: raw CT image with noise, segmentation of the raw CT image, segmentation of the denoised CT image

Fig. 1. A firefighter in full gear exits a fire. Firefighters often wear over 100 lb of gear and must work fires in brief shifts to avoid exhaustion.

When a responder goes into a scene, he only receives limited information from dispatch. Upon arrival to a scene to rescue a trapped victim, he may only know the number of floors the structure has. Upon entering the building, his vision is completely blocked by smoke. The search for victims is conducted blindly and, in higher temperatures, on hands and knees. Firefighters work in pairs, keeping one hand constantly on the wall or their partner while sweeping the floor with their other hand or one of the tools they have with them. Firefighters must also take in the location of doors and furniture to maintain their orientation. They must also keep track of time to ensure they don't run out of oxygen and clear the building before it becomes structurally unsound. If a firefighter is able to find the victim, he must now blindly recall his way out of a potentially collapsing building.

In order to respond to a scene, firefighters often have to carry heavy equipment such as oxygen tanks or medical devices. While on a scene, firefighters often have their hands occupied by tools such as a hook or axe. When considering new solutions for responders, it is critical for designers to be mindful of the physical limitations of what firefighters can carry in addition to their gear. Furthermore, responders are highly focused on the tasks they have to perform, so tools must be straightforward and require little extra physical or cognitive effort to use.

Research aimed to identify necessary tasks and information for firefighters to successfully complete their tasks. In early prototyping phases, researchers returned to fire stations at multiple points to validate that the information their concepts aimed to present was not only helpful but also acceptable to the target audience. An important discovery from these prototype validation sessions was that firefighters are willing to adopt technology that supplements their knowledge of the environment but resist technology that prescribes or suggests actions or decisions. Firefighters trust their own experience more than an algorithm and reject concepts that attempt to replace their decision-making abilities.

3 Our Solution

In order to address the constraints of the environment, researchers created quick augmented reality (AR) product concepts, which they rapidly prototyped and tested with users. Researchers then built a virtual reality (VR) environment to test concepts in concert with each other and the dynamics of emergency situations. The VR environment supported 3 tasks: Blind search, directed search, and exiting the building (Fig. 2).

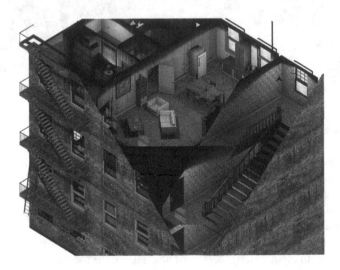

Fig. 2. A cut-out view of the testing environment.

VR Testing Environment. The environment involved both visual and audio factors to simulate real-world noise and stressors of a burning building, as captured by research. One of the major drawbacks of this environment was the tendency of immersive virtual reality environments to cause motion sickness in participants. In order to accommodate this, researchers limited the length of testing sessions to under 10 min and had users take frequent breaks.

The environment was a multi-story residential building. The rooms in the building contained furniture so that the simulation contained a realistic level of way finding. Interviews with firefighters revealed that they would map out the rooms in a building using the furniture they contained. In order to add noise to the visual channel, the rooms would fill up with smoke and visibility would decrease as the fire persisted. In order to add noise to the audio channel, the simulated fire made realistic sound and researchers piped radio chatter into the simulation.

A timing element was implemented in order to add stress to the task. The interface displayed a countdown clock that indicated when their oxygen would run out. As time went on, the size of the fire and the amount of smoke increased, adding urgency to the user's experience and hindering their ability to see and complete their task. Additionally, users were required to pay attention to and respond to certain cues from the radio clips.

The VR prototyping and testing environment was constructed using Unity3D and Oculus Rift. The elements of the Unity environment facilitated an iterative process wherein elements of the AR prototypes and the testing environment can be adjusted and changed quickly and independently. A virtual environment by nature captures performance metrics in extremely controlled and repeatable conditions, so designs could be compared using measures of the speed and accuracy of task completion.

Prototype. The prototype supported testing of 2 AR tools: a heads-up display (HUD) in combination with 3D audio (3DA) components. The HUD displays a layer of visual information to the user on top of the environment. 3DA uses stereo audio to mimic sounds originating from specific locations in 360° around the user. The prototypes tested how this additional information might be provided in a given environment as well as how it could assist a user in completing a simulated task. Prototypes were constructed to be modular and interchangeable for a variety of testing configurations. Users were also able to configure their own AR display combinations in later trials to create customized AR experiences.

Task: Blind Search. In order to support blind search, one prototype displayed a visual "tail" to indicate where the user had already searched. Researchers designed this interface to prevent disorientation or duplicate work. Initial testing revealed that haptic channels became overloaded and users reported becoming desensitized to and ignoring those cues. Early rounds of testing using haptic and audio prototypes indicated that the visual channel was the best fit for this information. Research showed users were confused when they received their search history over the audio channel, and this confusion persisted despite training sessions. The visual channel excelled in supporting this task once users voiced a desire to know the age of the trail. Testing revealed that color changes were the best way to display the age of the trail (Fig. 3).

Fig. 3. A screen capture from an initial prototype displaying the visual search history and the oxygen countdown (Color figure online).

Task: Directed Search. The next portion of the task involved a directed search for a fallen firefighter. Both the HUD and a 3D Audio tool proved usable for AR in this task. To assist directed search, the HUD displayed an arrow that pointed the user in the direction of the fallen teammate. An important distinction here was that the arrow was pointing in the absolute direction of the target and was not providing turn-by-turn directions through the building. The audio version of this AR used two different methods to guide the user toward their teammate. One method adjusted the repetition rate of a ping to be more rapid as the user was oriented toward the target. Since users wore an Oculus, the interface would pick up on their head orientation. This enabled users to turn their head, hear changes in the ping rate, and check that they were headed in the right direction. The second method adjusted the pitch of the ping to be higher as users approached the target and lower as they got further away. The prototype was designed to support swapping the pairings of repetition rate and pitch between distance and direction, which allowed users to customize their interface for this task according to what was most intuitive to them.

Task: Exit the Building. During initial concept validation, firefighters expressed concern that over-reliance on any additional technology would cause disorientation if the tech were to fail. The final task was designed to test user performance given AR device removal. Once users had found the victim, augmented reality displays were removed and users were required to navigate their way out of the building without any additional tech. This task measured whether the users were relying too heavily on the technology to still maintain a working knowledge of their spatial orientation. Ideally, AR will augment a user's ability to complete a task without becoming a crutch.

4   User-Centered Methods

Domain Research and Synthesis. Researchers began with an extensive literature review to gather domain knowledge of augmented cognition and firefighters. Researchers then performed structured interviews with 26 target end-users, domain experts, subject matter experts, and stakeholders.

After using interviews to gather domain knowledge, researchers gathered real world information through contextual design methods [1]. Researchers participated in 6 ride-alongs with target users to observe their workflow during emergency situations and attended a training session run by a fire fighter in order to understand the mindset, mental models, and rationale of emergency response.

Research methods had to be modified in order to accommodate the extremely dangerous conditions that users encounter. Traditional contextual design methods require closely following users during their tasks, but for safety concerns, the target user workspace did not support traditional contextual inquiry. In order to accommodate these dangers, researchers used a combination of observation and directed storytelling to have users recount a specific work experience. Researchers went to observe an emergency response and took note of which firefighter teams were there. Over the next 24 h, researchers were able to contact the firefighters observed on-scene and have them recount the experience while it was fresh in their memory. This allowed the research team to ask specific questions and get realistic answers from the target users without endangering themselves or distracting the firefighters from their dangerous work.

To synthesize findings, researchers performed a journey mapping exercise. Researchers took this map to the firefighters who were at the scene to have them give feedback, which revealed distinct phases of emergency response: pre-arrival, arrival, and scene response. Journey maps plot the flow of information and responsibility during task completion. This journey map focused on the experience of a single firefighting team. Researchers decided to focus on the scene-response phase, as the journey map revealed it exhibited the most communication breakdowns. This phase is characterized by triage, coordination, self-preservation, and high states of stress and physical exertion. Researchers decided to focus on the on-scene responders as target users, as tech already exists for scene overseers.

Initial Prototyping. In the first round of synthesis, researchers mapped out necessary on-scene data per firefighter feedback during interviews. This data was gridded in a matrix against situations in which users indicated they would have wanted more information. 17 of these data-situation pairings were translated into storyboards to procure user feedback. Storyboards are small, illustrated stories that describe a problem and a potential solution. They allow researchers to quickly and clearly communicate their concepts to target users to get feedback on the impact and feasibility of the concept. Research shows that presenting rapid concepts like this to users generates valuable and applicable design feedback [2]. Storyboard feedback directs researchers to probe how the tech would be accepted on the market. Feedback from storyboards allowed researchers to tighten the scope of their final design. Users were open to receiving additional information while on the scene, but they did not want a device suggest action – they did not trust an algorithm to make the right decision or to replace their "gut" instincts. Users were open to a tool that provided additional information to enhance their own experience and decision-making abilities.

Fig. 4. Two team members test out an early audio prototype to perform a directed search task

Fig. 5. A test user performs a search and rescue task in an early version of the digital prototype using the visual history trail HUD.

Fig. 6. Two members of the research team test a concept using a pre-existing digital environment to validate that haptic feedback can be incorporated into a digital interaction.

Once target scenarios were identified through the storyboarding exercise, researchers brainstormed the kinds of information displays that could improve the situation. Using a fail-fast mentality, the research team generated quick-and-dirty prototypes for each product concept to test whether or not each of these concepts would be interpreted clearly and improve the decision-making abilities of the user. By user-testing each prototype as early in development as possible, the research team ensured that the majority of development time was spent building a useful tool. These mid-level prototypes allowed the research team to hone the final design (Figs. 4, 5 and 6).

5 Conclusion

Emergency response scenes are tense, uncertain, and constantly evolving. Firefighting is characterized by the management of many unknown factors, forcing firefighters to constantly anticipate what could go wrong. Each firefighter needs to keep track of how a scene could evolve and to make the best split-second decision possible. Technology could assist firefighters in keeping track of their scene and reducing their cognitive load.

Firefighters are open to adopting new technology, as they are frustrated with the radio as the main on-scene communication device. Entering into an emergency situation with missing information, conflicting goals, and environmental pressure is no small task. Communication should be an asset in these situations, not a liability. Building a solution for firefighters would make a meaningful impact on public safety and could be applicable beyond the emergency response space.

The digital solution presented in this paper explores the value of delivering location information to firefighters. Team location is a critical piece of information for responders, enabling them to coordinate their actions, ensure the safety of themselves and one another, and speed rescue efforts. Location information is difficult to communicate over the radio, and layering audio and visual information over the responder's perception could assist in search and rescue tasks.

By using user-centered research and design methods, a team of human-computer interaction researchers quickly gained a deep understanding of the experience and challenges first responders face when responding to an emergency. They used this understanding to generate a multiple ideas for solutions, which they then represented as storyboards and prototypes and tested with their users. This let them quickly an inexpensively identify the concepts that users would find most useful. These concepts were built into prototypes of increasing fidelity, tested and iterated upon.

The researchers were also able to incorporate their background user research into the construction of a virtual reality testing environment. This let them test their prototypes in a realistic yet controlled environment in which they could gather rich data. The ease of creating prototypes in a digital environment also supported rapid prototyping and iteration. This, combined with a commitment to test with the end users throughout the process, ensured the delivery of a highly usable (product/prototype) at low cost and within a short timeframe.

References

1. Beyer, Hugh, Holtzblatt, Karen: Contextual Design: Defining Customer-Centered Systems. Morgan Laufmann Publishers Inc., San Francisco (1998)
2. Davidoff, S., Lee, M.K., Dey, A.K., Zimmerman, J.: Rapidly exploring application design through speed dating. In: Krumm, J., Abowd, G.D., Seneviratne, A., Strang, T. (eds.) UbiComp 2007. LNCS, vol. 4717, pp. 429–446. Springer, Heidelberg (2007)

RevealFlow: A Process Control Visualization Framework

Ronald Boring[1(✉)], Thomas Ulrich[2], and Roger Lew[2]

[1] Idaho National Laboratory, Idaho Falls, ID, USA
ronald.boring@inl.gov
[2] University of Idaho, Moscow, ID, USA
ulrich@uidaho.edu, rogerlew@vandals.uidaho.edu

Abstract. In this paper we describe current and historic permutations of control room technology and describe a new set of design principles for digitally displaying process control parameters. The design principles focus on helping operators effectively monitor changes during process control. The change detection approach is called RevealFlow and is illustrated in the context of the Computerized Operator Support System currently being developed for nuclear power plant control rooms.

Keywords: Process control · Distributed control system · Control room · Change detection · Computerized operator support system

1 Introduction

1.1 Generational Differences in Control Rooms

Industries like chemical, manufacturing, oil and gas, and energy involve multiple simultaneous processes. When multiple systems converge on a large scale, the process control facility may be said to be a plant, with designations as diverse as a chemical plant or a power plant. Typically each plant requires a control room as a central place to coordinate and control processes. While a control room may feature significant automation, operators still oversee the process from the control room, ensuring normal production and monitoring for anomalies, including threats to safety. Safety considerations become paramount, as a system malfunction can lead not only to equipment damage but also to harm to the environment or people at or near the plant.

Control room technology requires remote sensors and actuators, which rely primarily on electrical-mechanical components. While plants were possible without these technologies, the centralized control room was enabled with the advent of electrical gauges

This work of authorship was prepared as an account of work sponsored by an agency of the United States Government. Neither the United States Government, nor any agency thereof, nor any of their employees makes any warranty, express or implied, or assumes any legal liability or responsibility for the accuracy, completeness, or usefulness of any information, apparatus, product, or process disclosed, or represents that its use would not infringe privately-owned rights.

© Springer International Publishing Switzerland 2016
D.D. Schmorrow and C.M. Fidopiastis (Eds.): AC 2016, Part II, LNAI 9744, pp. 145–156, 2016.
DOI: 10.1007/978-3-319-39952-2_15

and switches in the 1920s [1]. Large ships are good examples of the emergence of control rooms. Steamboats brought the separation of engine room below deck and the bridge above deck. The captain or pilot set the speed and direction of the engine using the engine order telegraph, in which the captain's setting was mirrored in the engine room. Changes to the dialed position were accompanied by audible bells in the engine room to alert the engineers that they needed to change the engine speed or direction. Status indications between the engine room and bridge were also possible through telegraph, telephone, or intercom. As remote sensors and remote-controlled switches became available, the bridge was equipped with gauges to allow direct monitoring and provide direct control over the engine or other facets of the ship. The role of the ship's engineer shifted from that of control and maintenance of the engine to primarily maintenance of the engine. The control room eliminated the need for redundant personnel to relay status or control information.

1.2 Analog Control Rooms

Beginning in the 1940s, analog control rooms began to take root. The term analog is used to describe the human-system interaction used by the operators and may not necessarily apply to the technologies behind the board. Several standard characteristics of the centralized control room emerged. These included:

- *One-for-one arrangement.* In traditional analog control rooms, each instrument or indicator is directly wired to an equivalent sensor, and each control is directly wired to an actuator in the plant. There are no shared conduits or channels of information, and there is no aggregation of information or controls.
- *Simple indicators.* These indicators provide information about a single parameter like pressure level, flow rate, or temperature. Alternately, they may represent simple on-off logic like the status of charging pump or an alarm setpoint. The defining characteristic of these indicators is that they do not combine information from multiple sensors that would require computational logic or mathematical functions. Operators must integrate multiple indicators to assess the state of the plant.
- *Stand-at-the-boards operation.* While simple control boards were possible from a seated position, as additional instrumentation and controls (I&C) became available, it became necessary to expand the real estate of the boards vertically upward and horizontally outward. This arrangement eventually necessitated standing for some operations. The placement of some instrumentation higher vertically allowed monitoring supervision from across the control room.
- *Triple-layer design.* As noted, the control boards grew from operation for a seated to a standing position. A standard control layout evolved from this practice in which controls tended to be mounted low on the boards, often in a desk-like horizontal benchboard configuration. Above the desktop, a vertical panel comprises the second layer containing key instrumentation required for monitoring and control decisions. Finally, higher up the boards were found alarm lights. In this manner, immediately required information was close to eye level of the standing operator, and controls

were within arm's reach. Information such as alarms, needed only at the level of catching the operator's attention, was placed high on the boards.

- *Setpoint alarms.* With remote sensors came the technology for setpoint alarms. These alarms were triggered when a particular measured entity reached a particular threshold, e.g., when a pipe exceeded the maximum recommended operating pressure. The threshold setpoint activated a light in the control room, which either contained a label near it or was placed in a lightbox with illuminated text upon activation. Additional features like audible alarms, flashing alarms, and silence buttons were added to the configuration, but the alarms continued to be based on the simple threshold setpoints.

- *Simple controls.* These controls are tied to a single function, usually equivalent to an on-off switch to activate a motor that in turn opens or closes a valve or pumps fluids. Typically, a control does not activate a series of sequential controls nor perform simultaneous parallel control actions. These simple controls may feature electrical or mechanical lockouts to prevent erroneous activation (e.g., turning on a pump to remove fluid when another pump is injecting fluid), and they may feature auto-stop for when a particular state (e.g., full valve open) is achieved. The controls may also feature two-factor confirmation such as when two buttons are required to be pressed simultaneously to close the circuit for emergency shutdown. In the latter case, because the consequences are high (e.g., cost of lost production or potential loss of equipment by sudden shutdown), the lockout serves to safeguard against inadvertent activation.

- *Manual operation.* Mechanical safety actuations like pressure relief valves, shear points, and electrical fuses were possible, but the control room did not feature automation. The plant was controlled entirely by the operator. A characteristic of much of process control is the achievement of steady state operations, which require minimal adjustment by the operator. However, plant transients might require extensive adjustments in prescribed sequences.

- *Procedures.* While procedures may not be part of the physical characteristics of the control room, they were increasingly required to support operations and maintain the plant within a known safety envelope, especially during transient conditions where the sequence or prioritization of particular actions was important. Eventually, e.g., in nuclear power plant control rooms, procedures became such an integrated part of the control room that special places were set aside to house the procedures within or around the control panels.

- *Command-and-control crew operation.* As the complexity of the plant process grows, the need for multiple operators likewise increases. As such, complex plants often required more than one operator. When there are multiple operators, there may be a supervisor to orchestrate actions and maintain process overview while operators monitor and control subsystems. Thus, while an individual operator may be involved in the minutia of controlling one particular system, the supervisor maintains situation awareness for the overall process. In some arrangements, the supervisor may also be in charge of issuing directives to the operators, establishing a command-and-control arrangement. Many plants have adopted a threeway communication protocol in which the supervisor issues a command or request, the operator repeats it back, and the supervisor confirms the operator has correctly understood the communication.

These features are not mutually exclusive, nor are they a template that is found in all analog control rooms. They simply serve as a reference set of features commonly observed in analog control rooms.

1.3 Digital Control Rooms

The introduction of digital technologies to the human-system interfaces within control rooms has fundamentally changed the features and functions of control rooms. Digital control rooms may feature [2]:

- *Multipurpose displays and soft controls.* A distributed control system (DCS) features one or more displays with input capability such as a mouse, trackpad, or touchscreen [3]. These displays may, in the architecture of the DCS, be toggled between different system function screens. As such, it is not necessary to have all information displayed simultaneously across the boards, as a single display can present distal information at one physical location. The input device likewise features remote control from a single location by providing virtual or soft controls tied to the particular screen on the display.
- *Information integrative indicators.* Automation may take the form of information automation and control automation. Information automation combines disparate information that operators would otherwise have to gather and assemble to draw a conclusion or maintain overview. With the highly distributed nature of information in analog control rooms, operators often needed to ping-pong back and forth to maintain situation awareness of processes. Digital displays can consolidate information that would otherwise be widely dispersed across the boards. Moreover, digital displays can provide aggregate views that support the operators, e.g., custom trend displays of key parameters or calculations of composite measures (e.g., overall loss of cooling rate given a failed cooling water pump) that would normally be performed manually by operators or technical support staff in the control room.
- *Complex or automated controls.* As noted, digital controls no longer require a physical switch on the control boards, as they can be controlled remotely through the DCS using soft controls available on the screens dedicated to each system in the plant. The control functions do not need to be linked to a single action, and it is possible to combine a chain of actions for each soft control. In some plants, for example, it is possible to have single-button startup or shutdown sequences without the need for ongoing human intervention. These features are a form of automation; it is also possible to have full automation for large facets of plant operations.
- *Console or workstation operation.* Digital control rooms often forgo panels because of the space efficiency and convenience of consolidating I&C on the DCS displays. With the advent of DCS workstations, the need to stand at the boards is diminished, and the workstations are often designed for seated operators. Some backup panels may be retained for safety in the event of DCS failure, but most DCS architectures feature redundant hardware and the ability to pull up any screens from any display.

Thus, in the event of failure of one operator workstation, the operator could simply go to a backup workstation and resume the full range of process control for the plant.

- *Overview displays.* The triple-layer design of analog control boards is no longer required when most monitoring and control take place from a desk. However, because digital monitoring information is localized to the individual operator, it is desirable to have a shared frame of reference in the control room. Overview displays, in particular large overview displays [4], provide a way to monitor overall plant status that may not be possible with system-specific screens. The overview displays also enable troubleshooting between operators and supervisors in the control room by ensuring all parties have the same visual information during group discussions. Overview displays do not generally allow control actions and therefore serve only the function of providing visual indicators to aid operators.

- *Advanced alarm systems.* By adding control logic beyond the simple alarm thresholds found in analog alarms, it is possible to add significant functionality to alarms. For example, it is possible to exercise state dependence, by which only alarms relevant to a particular mode of operation are enabled. This feature overcomes the problem of alarms for steady state operations activating during startup or shutdown. Further, it is possible to implement alarm grouping, such that only a single alarm activates when a whole group of interrelated alarms might activate otherwise. Because process control often involves a sequence of activities, failure in one part causes a cascade of failures and corresponding alarms, which can result in an alarm flood that obscures the root fault. Other advanced alarm features include prioritized alarms that indicate severity to allow operators to take quick action in the event of multiple faults, prognostic and predictive alarms that anticipate faults, and advanced visual alarms that depict the fault in such a manner that the operator is able unambiguously to see the fault in context.

- *Single operator control.* Whereas analog control rooms often required multiple operators performing actions under direction of a supervisor, DCS technology provides the operator with the ability to perform actions independently. Features such as computer-based procedures eliminate the need for a supervisor to coordinate procedures. Additional features like automation reduce the need for constant operator vigilance and may reduce the need for multiple operators. Thus, advanced digital control rooms often yield a greatly reduced crew complement, sometimes resulting in only a single operator to oversee a large plant.

As with analog control rooms, it must be noted there is no prototypical digital control room, and different features will likely be present for each particular implementation. An important consideration for digital control rooms is that they chronologically are newer and have benefitted from the nascence of human factors engineering applications in control rooms [5]. Human factors has resulted in improved design to the flow of activities in the control room, presentation of information to operators, and workflow of the operators. The marriage of automation technology, advanced visualization capabilities, and human factors optimization have resulted in significantly improved control rooms compared to their predecessors.

1.4 Control Rooms in U.S. Nuclear Power Plants

Idaho National Laboratory (INL) is engaged in human factors research in support of control rooms for the U.S. energy sector. Much of this work centers on nuclear power plant applications, where there is a twofold mission to modernize the control rooms of existing plants [6] and to develop new control room concepts for advanced reactor designs like small modular reactors [7].

The existing U.S. fleet of commercial nuclear power reactors is aging, and many plants are drawing to the end of their original 40-year operating license. While some utilities have chosen not to extend the license of a plant and commence decommissioning, in the vast majority of cases, the utilities that operate the plants are choosing to apply to the U.S. Nuclear Regulatory Commission to extend the operating license by another twenty years. The initial operating period was fully anticipated, and utilities stockpiled replacement parts to ensure safe and reliable operation. Replacing worn or broken components with equivalent components also ensured that the plants successfully operated within their original licensing basis without potentially requiring license amendments to accommodate the introduction of new technology. With license extensions, the plant may find itself nearing the end of useful life for existing equipment or at the point where the cost of refurbishment or like-for-like replacement parts exceeds the cost of new equipment. At this point, the utility is confronted with the unique problem of finding new equipment that serves the same function as existing equipment and determining if the new equipment fundamentally changes the conduct of plant operations such that a license amendment might be required.

INL supports efforts to modernize nuclear power plant main control rooms, featuring a stepwise, system-by-system upgrade path [8]. This path results in a hybrid control room consisting of a mix of analog-mechanical and digital I&C. Although the term *digital island* is sometimes used pejoratively to describe the introduction of limited digital systems into existing analog control rooms, the first DCSs introduced to the control boards are an important stepping stone toward fully digital control rooms. The feasibility of performing a large-scale control room replacement is explored in [9], and nuclear utilities indicate that they are unlikely to be able to replace the entire control room at one time due to loss of revenue during the extended outage required for such a control room replacement [10]. Instead, the utility undertakes a gradual upgrade process, typically consisting of one system or board per refueling outage. INL has designed the Guideline for Operational Nuclear Usability and Knowledge Elicitation (GONUKE) [11] to provide a process suitable for design and evaluation of new digital systems that are introduced to the control boards.

An analogous design transformation can be seen in commercial airplane cockpits, which have seen the significant introduction of new digital controls. Initial efforts resulted in the insertion of retrofitted multifunction displays into the cockpit to replace existing analog I&C. In most cases, the multifunction displays added avionics functionality to aid the pilot, from digital pitch and roll data, to navigation functions, to weather and airspace, to autopilot, to collision avoidance systems. Retrofitted cockpits offer different levels of digitization, from hybrid avionics to completely digital glass cockpits.

Control rooms for new nuclear power plants subscribe to many of the features indicted in Sect. 1.3 of this paper. There exist some regulatory barriers to full adoption of all features found in other industries. For example, the heavy emphasis on safety has resulted in the requirement to maintain crew staffing levels analogous to analog control rooms. Additionally, the need for transparency in control logic has resulted in minimal intelligent or autonomous control. Examples of three generations of nuclear control rooms are depicted in Fig. 1.

Fig. 1. Three generations of nuclear power plant control rooms (top to bottom: EBR-1, the first nuclear power plant with an all analog control room; recently decommissioned San Onofre Nuclear Generating Station, with a hybrid analog-digital control room; HAMMLab at Halden Reactor Project, a fully digital advanced control room concept).

2 The Need for New Visualization in Control Rooms

The previous sections provide extensive background on the different types of control rooms. Conventional analog control rooms, such as those commonly found in nuclear power plants, represent information in a parallel fashion, typically with a one-to-one mapping of sensors to indicators. This design approach requires extensive control room real estate, especially for complex control system processes. As digital control systems, such as those found in modern control rooms for electrical grids or gas distribution networks, have begun to replace analog I&C, they have afforded the opportunity to use common displays across all systems, thereby providing a smaller footprint in the control room. The approach often uses a nested navigation scheme, whereby control operators have on-screen windows for particular subsystems.

Both approaches represent tradeoffs. For analog control rooms, operators must scan across control panels to maintain their plant overview, a complex process that demands the operators to integrate and track multiple simultaneous indicators. This disadvantage is offset by the ability to see all information at once, thereby minimizing the danger that critical indicators will be hidden in nested windows. In contrast, for digital control systems, the operators are able to avail themselves of optimized displays, including key parameter displays. However, having information consolidated on single windows may result in loss of situation awareness by these operators, as critical windows must often be toggled back and forth, thereby reducing the overview the operator may have of the larger process being controlled.

The shift to digital control rooms is inevitable, whether performed as a stepwise upgrade process or as a complete control room replacement. Successful deployment of digital technology in control rooms requires effective ways to display crucial indicator information to operators in order to allow them to monitor plant status and diagnose problems. To combat the loss of situation awareness inherent in nested displays in process control, designers of DCSs have developed overviews, often displayed as large overview displays, viewable by multiple operators across the control room. The challenge with such displays is they do not inherently reduce the problem of information overload that confronts the operator of a complex system. Design techniques for representing information in an intuitive manner help to reduce the workload in processing key information, but they do not necessarily reduce the overall amount of information the operator must monitor and process in parallel. The danger is that the operator may miss an important change in a key parameter because of the large number of visible indicators. If such is the case, eventually an alarm will indicate once the parameter moves out of acceptable bounds, but this alarm may come only at the point when remediation is necessary. Thus, the key operator role of monitoring and preventing upsets is not realized.

Several design philosophies have been created for control room visualizations, including ecological interface design (EID) [12, 13], information rich design (IRD) [14], and high performance human-machine interface (HMI) principles [15].

- EID is a design approach that strives to present the operational constraints in a natural manner for key process parameters. This approach specifically capitalizes on the complex interactions inherent in process control systems by focusing on how to provide operators with sufficient context embedded within a parameter to understand what that parameter is doing and determine where the safe operating bounds are for that parameter.
- IRD aims to create high information density displays without overloading the operator. The basic design concept consists of muted or so-called dullscreen displays in which only important information is made salient through color. This approach is optimized for process control in that it allows a large number of process variables to be displayed concurrently.
- Finally, there is high performance HMI. Both EID and IRD produce uniquely identifiable displays. High performance HMI is not so much a single set of design principles as it is a process to infuse a systematic design across the control room.

The key elements are adopting a style guide based on human factors principles and deploying that style guide consistently to design or redesign the control room.

EID, IRD, and high performance HMI are not incompatible approaches, and it is possible to use elements of all three approaches in concert. These approaches have yielded effective digital control rooms, but they represent a very small set of the possibilities for control room design. In the remainder of this paper, we present a novel approach to visualizing process control indicators.

3   RevealFlow

3.1   Design for Change Detection

Change blindness [16] occurs when people fail to detect a change in visual stimuli. Change blindness may occur when changes in the visual stimuli are not salient enough to detect, but it may also occur even when the changes are sufficiently salient. A person may be focused elsewhere during key changes, a person may undertake eye movement (i.e., visual saccade) during changes, a person may be overloaded in terms of the number of items concurrently attending, or a person may simply experience perceptual overload. Change blindness and the related concept of inattentional blindness are regarded as sources of error in control room operations [17]. For example, an operator may miss a key plant indicator because he or she is not attending to that part of the boards. In an analog control room, where there may be limited trend displays, the change may, in fact, not be obvious until it reaches a critical level such as an alarm state. Periodic surveillance of key indicators to expected levels helps to minimize the opportunity for such initial misses to generate any consequence to the plant, but such surveillance does not guarantee catching missed changes in parameters.

One of the major tasks of operators is to monitor and detect changes to the plant process. In the case of intended transients to the plant like startup or shutdown, the operator ensures that parameters change at the expected rates and to the expected magnitudes. A key indicator like differential rotor temperature that is too high during startup, for example, has the potential to damage turbomachinery. As another example, a flow rate that suddenly changes in an unexpected manner could be indicative of a leak or blockage in the system. A safety alarm will notify when levels are beyond specified setpoint thresholds, but the operator can play a crucial role in early detection of anomalies in the process. When an operator recognizes a trend toward an anomaly, he or she can intervene before the fault becomes a serious threat to safety, the plant, or the process.

Curiously, despite the centrality of change detection to operators, process control solutions have neither reliably nor effectively helped highlight changes to plant indicators. A proactive display strategy is necessary to help operators maintain process oversight while detecting key changes in indicators. Here, we introduce the RevealFlow visualization framework, an approach that accentuates the operators' ability to detect changes in the process. RevealFlow consists of four guiding principles:

1. *For process monitoring, changes are equally important to steady states.* Most control room indicators provide the current state in a numeric depiction, either as an

alphanumeric value or as a graphical representation of that value. However, it is often not the current magnitude of the indicator but rather the change in the magnitude that is of interest to the operator.

2. *Changes should be apparent at all times.* A plant undergoes fluctuations, and these dynamics are important for the operator to be able to see at all times. Changes should be highlighted on the display in a manner that allows them to stand out from steady state values. Small fluctuations within a deadband should be ignored. Larger changes should be visible with their salience proportional to the magnitude of the change. These changes, once highlighted, should remain prominent.

3. *Changes occur in historical context.* To represent change, it is important to show how the indicator has changed over a period of time. Essentially, changes must be trended visually. Note that trending a change is slightly different than trending an indicator's value. Change trending captures the derivative of the dynamics vs. simply the history of the indicator.

4. *Changes escalate, but their cause does not.* A complex process may feature many interconnected systems that can result in a chain reaction when one part of the process is disrupted. Similar to an alarm flood, it is possible that these changes, when simultaneously active, may overwhelm the operator's ability to detect the most important changes. Process control is "big data," and preventing information overload requires process information filtering. As such, RevealFlow recommends emphasizing first-out changes and those changes that are of highest priority to the plant.

It should be noted that these principles represent a design philosophy, not a prescriptive design style guide. A translation of these principles into example designs for process control systems is provided in the next section.

3.2   Examples of RevealFlow

RevealFlow is being implemented in the Computerized Operator Support System (COSS) [18], a digital operator aid that provides ongoing process monitoring in control rooms. Currently, COSS is installed in the Human Systems Simulation Laboratory [19], a full-scope nuclear power plant control room simulator facility at INL. COSS consists of a DCS with an advanced HMI frontend and intelligent process diagnosis (PRODIAG) system backend [20]. PRODIAG acts as a detection engine to determine changes to modeled parameters. Confluence equations serve to provide unit-neutral metrics of change ideally suited to process monitoring. RevealFlow represents an effort on behalf the authors to provide a usable visual representation in the COSS HMI for the change monitoring in PRODIAG.

The visualization scheme borrows from EID and IRD. A simple example is provided in Fig. 2. As currently envisioned in COSS, RevealFlow begins with a dull-screen visual outline of key parameters. Light-colored indices are displayed as grey graphs arranged along a functional piping and instrumentation outline. As indicators change, they are highlighted on the graphs such that they become readily visible to the operators. A greyscale gradation to the steady state of the indicator allows the operator to see the change over time, while a color highlight indicates an alarm state.

RevealFlow is able to address the issue of large data visualization by only high-lighting changes. When integrated across an overview display, the RevealFlow graph-ical elements create a muted backdrop for in-range indicators, only drawing attention to important changes. Gradual, slow drift changes are greyscale and less salient, while rapid shifts are highlighted with color.

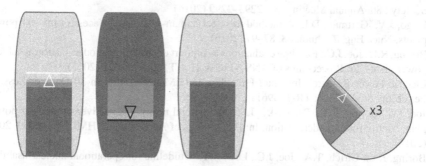

Fig. 2. Three examples of RevealFlow bar graphs indicating an elevated state relative to the setpoint (left), a low alarm state (middle), and a normal state (right). The greyscale gradation allows a temporal trail, which represents time information trending not typically represented outside line graphs. An arrow overlaid on the graph draws attention to changing indicators. On the far right, there is a pie chart illustrating the concurrent change in three related systems.

4   Conclusions

Existing control rooms—whether analog, hybrid, or digital—represent strategies for capturing plant process data for monitoring by the plant operators. Rarely do these strat-egies help operators focus on arguably the most important aspect of the plant processes, namely those processes that are experiencing change, especially those with rapid shifts. RevealFlow presents a design approach centered on simplifying the display of infor-mation to highlight process dynamics rather than process states. RevealFlow is being implemented in COSS, and new graphical visualizations are ongoing. The culmination of RevealFlow will be the evaluation of the effectiveness of the RevealFlow designs benchmarked against other control room technologies. It is hypothesized that Reveal-Flow will simplify monitoring of complex processes. RevealFlow therefore represents a significant shift in concept of operations and will require extensive evaluation to ensure the efficacy of the design principles. Change is constant in process control; perhaps RevealFlow will prove a worthy change to control rooms.

References

1. Bennett, S.: A History of Control Engineering, 1930–1955. Peter Peregrinus Ltd., London (1993)
2. Furet, J.: New Concepts in Control-Room Design. IAEA Bulletin (Autumn 1985)
3. Ulrich, T.A., Boring, R.L., Lew, R.: Control board digital interface input devices—touchscreen, trackpad, or mouse? In: Resilience Week Proceedings, pp. 168–173 (2015)

4. Jokstad, H., Boring, R.: Bridging the gap: adapting advanced display technologies for use in hybrid control rooms. In: Proceedings of ANS NPIC & HMIT, pp. 535–544 (2015)
5. Strobhar, D.A.: Human Factors in Process Plant Operation. Momentum Press, New York City (2013)
6. Boring, R.L.: Human factors design, verification, and validation for two types of control room upgrades at a nuclear power plant. In: Proceedings of the Human Factors and Ergonomics Society 58th Annual Meeting, pp. 2295–2299 (2014)
7. Hugo, J.V., Gertman, D.I.: A method to select human-system interfaces for nuclear power plants. Nucl. Eng. Technol. **48**, 87–97 (2016)
8. Boring, R.L., Joe, J.C.: Baseline evaluations to support control room modernization at nuclear power plants. In: Proceedings of ANS NPIC & HMIT, pp. 911–922 (2015)
9. Electric Power Research Institute: Full Plant I&C Modernization in 30 Days or Less, A Feasibility Study, EPRI TR-1009611 (2004)
10. Joe, J.C., Boring, R.L., Persensky, J.J.: Commercial utility perspectives on nuclear power plant control room modernization. In: Proceedings of ANS NPIC & HMIT, pp. 2039–2046 (2012)
11. Boring, R.L., Ulrich, T.A., Joe, J.C., Lew, R.T.: Guideline for operational nuclear usability and knowledge elicitations (GONUKE). Procedia Manuf. **3**, 1327–1334 (2015)
12. Vicente, K., Rasmussen, J.: Ecological interface design: theoretical foundations. IEEE Trans. Syst. Man Cybern. **22**, 589–606 (1992)
13. Vicente, K.: Ecological interface design: progress and challenges. Hum. Factors **44**, 62–78 (2002)
14. Braseth, A: Information-rich design for large-screen displays. Nucl. Eng. Int. Mag. 22–24 (2014)
15. Hollifield, B., Oliver, D., Nimmo, I., Habibi, E.: The high performance HMI handbook: a comprehensive guide to designing. In: Implementing and Maintaining Effective HMIs for Industrial Plant Operations. PAS, Houston (2008)
16. Beck, M.R., Levin, D.T., Angelone, B.: Change blindness blindness: beliefs about the roles of intention and scene complexity in change detection. Conscious. Cogn. Int. J. **16**, 31–51 (2007)
17. Whaley, A.M., Xing, J., Boring, R.L., Hendrickson, S.M.L., Joe, J.C., Le Blanc, K.L., Morrow, S.L.: Cognitive Basis for Human Reliability Analysis, NUREG-2114. U.S. Nuclear Regulatory Commission, Washington, DC (2016)
18. Boring, R.L., Thomas, K.D., Ulrich, T.A., Lew, R.T.: Computerized operator support systems to aid decision making in nuclear power plants. Procedia Manuf. **3**, 5261–5268 (2015)
19. Boring, R.L. Overview of a reconfigurable simulator for main control room upgrades in nuclear power plants. In: Proceedings of the Human Factors and Ergonomics Society 56th Annual Meeting, pp. 2050–2054 (2012)
20. Villim, R.B., Park, Y.S., Heifetz, A., Pu, W., Passerini, S., Grelle, A.: Monitoring and diagnosis of equipment faults. Nucl. Eng. Int. Mag. 24–27 (2013)

Paradigm Development for Identifying and Validating Indicators of Trust in Automation in the Operational Environment of Human Automation Integration

Kim Drnec[✉] and Jason S. Metcalfe

Army Research Laboratory, Aberdeen, MD, USA
kim.a.drnec2.ctr@mail.mil

Abstract. Calibrated trust in an automation is a key factor supporting full integration of the human user into human automation integrated systems. True integration is a requirement if system performance is to meet expectations. Trust in automation (TiA) has been studied using surveys, but thus far no valid, objective indicators of TiA exist. Further, these studies have been conducted in tightly controlled laboratory environments and therefore do not necessarily translate into real world applications that might improve joint system performance. Through a literature review, constraints on an operational paradigm aimed at developing indicators of TiA were established. Our goal in this paper was to develop an operational paradigm designed to develop valid TiA indicators using methods from human factors and cognitive neuroscience. The operational environment chosen was driving automation because most adults are familiar with the task and its consequent structure and therefore required little training. Initial behavioral and survey data confirm that the design constraints were met. We therefore believe that our paradigm provides a valid means of performing operational experiments aimed at further understanding TiA and its psychophysiological underpinnings.

Keywords: Trust in automation · Operational paradigm · Driving automation · Human automation integrated systems

1 Introduction

Joint human automation systems have been developed to leverage the abilities of both agents in order to improve overall task performance. However, true integration has yet to be realized, and the automated agent is often either misused, or disused entirely resulting in relatively poor performance outcomes. One reason genuine integration has not yet been achieved is an apparent lack of user acceptance. The degree to which a human user accepts an automated agent is thought to be directly related to the level of trust the human user has in the automation [1–3]. That is, as people gain confidence in the reliability, robustness, and safety of automated technologies, they develop sufficient trust to willingly share important decision and/or control authority with such systems. Therefore, if automated systems are to be used as designed, enabling joint system performance to reach intended levels, it is important that the human user develop a certain level of trust in the automation (TiA). However, more important than achieving

© Springer International Publishing Switzerland 2016
D.D. Schmorrow and C.M. Fidopiastis (Eds.): AC 2016, Part II, LNAI 9744, pp. 157–167, 2016.
DOI: 10.1007/978-3-319-39952-2_16

a certain level of TiA is to manage it so that behavioral outcomes such as misuse or disuse [6–9] do not occur regularly and negatively impact overall performance. Consequently, an important goal for systems designers is to find a means to calibrate the human user's TiA to elicit desired interaction with the automation given the nature of the ongoing, and dynamic, task context [3, 10–13]. An immediate need if TiA is to be calibrated is to establish quantitative and easily monitored indicators of TiA that are robust across individuals, task and time. Currently the only method for assessing TiA is by participant self-report through survey instruments, and few of these surveys have been validated. However, if objective real-time measurements of TiA can be identified and demonstrated as valid, systems could be designed to measure and manage TiA for real world applications that would maximize joint system performance of critical human automation system tasks.

The goal of this paper is to discuss the conceptual underpinnings of an operational research paradigm aimed at inference and validation of TiA outcome measures. First, we discuss important design constraints to such a paradigm based on human factors research. We then provide an example of how these concepts were realized in operational research that adapts methods from cognitive neuroscience and human factors engineering for addressing important issues for TiA and its influence on human-automation systems. Finally, we provide preliminary high-level analysis of an instantiation of our proposed paradigm demonstrating that it meets design constraints, and is therefore suitable as a method to identify indicators of TiA.

2 Concepts for Applying Cognitive Neuroscience to the Operational Study of TiA

Although methods from cognitive neuroscience have been applied in experimental settings to adaptive human automation systems that scale or mitigate task demands on the human user [4, 5] there has been little consistency in how the methods have been applied to specifically study TiA. We propose that such methods, particularly those based in psychophysiology, have considerable potential to effectively identify indicators of TiA if applied under appropriate operational constraints. The basis of this proposal is the understanding that trust is a psychological construct and therefore it would seem reasonable that there would be dynamic psychophysiological variables that enable inferences regarding extant levels of TiA for a given human user. Indeed, research on interpersonal trust has revealed measurable physiological changes correlated with changing participant trust and trust based decision making [6]. Therefore we believe that the application of these cognitive-neuroscience methods is promising for the study of TiA. However, much research across these domains (both cognitive neuroscience and human factors) has been laboratory based, leveraging dramatically simplified tasks performed in controlled environments, often using a narrow set of psychophysiological and/or behavioral data. These methods have resulted in important insights about cognitive and behavioral phenomena underlying human-automation relationships, but these laboratory-based research findings may be of limited value in more complex operational contexts. This is because they tend to apply to general populations rather than providing an understanding of how human-automation relationships develop as individuals perform tasks with real-world risks and consequences. New research paradigms are therefore required if an understanding of individual relationships are to be understood and leveraged to measure and manage TiA dynamics for particular operational environments.

3   Design Constraints

In order for research in this domain to be of use in operational settings, it is important that experimental conditions engender, as close as is reasonable, authentic levels of trust in ways reflective of operational influences. The relevant literature suggests three critical design considerations if this goal is to be met, (1) establishing a task-relevant risk and consequence structure, (2) engendering TiA levels as a function of automation relia-bility, and (3) engendering TiA levels as a function of workload. In addition to the theoretically based design constraints, it is critical, if human automation interaction is to be studied, that the subject be motivated to use the automation in a way that is organic to the operational environment. Moreover, we argue that it is critical to develop and validate that these factors have been successfully implemented if the paradigm is to be useful for more detailed research in the cognitive and neural underpinnings of variations in TiA and TiA-related decisions with regard to interactions with automation.

3.1   Risk and Consequence

Development of TiA requires inducing the perception of task-related risk or conse-quence to the human user [7, 8]; if consequences are low or irrelevant to the human, levels of TiA fail to be important. Generally speaking, we consider that without risk, trust is irrelevant to decision making. In order to develop a sense of risk and consequence it thus appears necessary to facilitate a sense of personal investment in the task outcome. While there may be multiple ways to achieve this, one of the more common methods in research has been to link performance outcomes with extrinsic rewards. Typically, these rewards are financial because most adults have daily experience with financial motiva-tion or gain. Though not directly applicable to many operational contexts, we chose financial motivation as a proven means of creating the needed senses of task investment and risk. Certainly, given the high cost of vehicle-based incidents, financial concerns tend to be common among real-world drivers as well.

3.2   Engendering TiA as a Function of Automation Reliability

Research has yielded much evidence as to what intrinsic and external factors affect extant levels and dynamic changes of TiA in the operational context of human automation inte-grated systems [1, 8–10]. In the general case, the degree or level of TiA appears to result from the evaluation of observations against *a priori* expectations about how an automa-tion should behave. Initially, most people would expect a real-world automation to be reliable and to be consistent over time, as well as being able to aid in achieving the task goal [1, 11]. Thus, with some exceptions [11], most human users will have an *a priori* expectation that the automation will be trustworthy, and therefore the initial level of TiA is likely to be relatively high. Reliability, or the degree to which the human user perceives the automation to be accurately performing tasks for which it was designed has significant effects on TiA levels is especially important at the start of automation use. Subsequently, consistency over time becomes critical to dynamic patterns of TiA levels; human users will continue to use, and even benefit from a slightly unreliable (above 70 % reliable) automation if the errors are predictable and consistent over time [12, 13].

3.3 Engendering TiA as a Function of Workload

Workload has been well established as a key influence on behaviors that have traditionally been attributed to TiA. For instance, under high-workload conditions, some people will choose to use an automation for which they hold low trust simply because some assistance is presumed to be better than none [14, 15]. Conversely, research also suggests that under conditions of low workload, human users tend towards manual operating mode [14, 16], likely because of boredom [16]. Therefore, this and other previous research has clearly demonstrated the interaction between workload and trust, leading to the expectation that the effects on psychophysiological variables for each factor would be difficult to disentangle. An operational paradigm focused on real-world outcomes should thus carefully consider the impact of workload in the design of their study on TiA.

3.4 Motivation

If the automation is never used, TiA levels cannot be established, and further, there is no interaction to observe. However, if the participants are rewarded or otherwise explicitly instructed to use the automation, results may reflect experimental design rather than the influence of TiA. One way of motivating natural interaction with the automation is to introduce automation independent secondary task of high value; the logic underlying this is that an automation that sufficiently handles lower value task elements will free operator resources to handle the higher value task. For instance, while modern driving automations are designed to prevent vehicle-vehicle collisions, not all are as capable of predicting and responding to the sometimes erratic and suddenly changing behavior of pedestrians (and other drivers). Therefore, it would be an appropriate driving strategy to engage a driving automation to manage vehicle control, enabling the human occupant to remain vigilant for pedestrians and similar potential hazards.

4   Implementing Operational Constraints into a Research Paradigm

Consider the example of our recently developed leader-follower driving paradigm during which participants were asked to perform a set of tasks relevant to real-world driving. Driving is a model paradigm for our purposes for several reasons. Driving is a task that many people engage in daily, and therefore little training is needed for subjects to perform an experimental driving task. In addition, driving automations are becoming increasingly common and consequently people are interacting with automations in a natural way. Therefore, an experimental driving paradigm appears to be an excellent operational context to address our questions regarding TiA and human automation interaction.

4.1 Primary Task and Environment

Participants were instructed to drive a simulated vehicle one full lap around a two-lane course. Task objectives included lane position control and maintenance of a "safe" distance from other vehicles, and particularly the lead vehicle in front of them. Automations with different capabilities were presented in different experimental conditions. For conditions in which the automation was available, participants had the option to enable or disable the automation at any moment. Lateral (wind gusts) and longitudinal (lead vehicle speed changes) perturbations were introduced to further challenge the performance of the driving task. In addition, participants were solely responsible for avoiding collisions, as the automation had no explicit collision avoidance capabilities. Therefore, the chosen automation independent secondary task involved avoiding collisions with frequently-appearing pedestrians by responding to them with button presses on a game controller. Pedestrians appeared approximately once every 6 s, distributed randomly on either side of the road, and 15 % stepped in the vehicle path.

4.2 Risk and Consequence

Risk and consequence were expressly manipulated through use of a game-like scenario where each deviation from task parameters had a preset consequence that was known to the participants. The point structure was chosen to encourage a specific hierarchical economy of decision making that was reflective of the risk structure in the real world. For example, collision with a pedestrian incurred the most severe penalty, whereas an

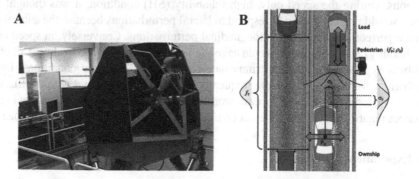

Fig. 1. Summary of experimental paradigm. (A) Ride Motion Simulator shown as a participant completes the driving task while wearing a 64-channel EEG cap. (B) Experimental task. Subjects drove a vehicle (ownship; right lane follow) while following a lead vehicle (right lane lead) and were instructed to maintain following distance and lane position. The varying reliability (low and high) of the driving automation are represented by the distributions labeled σ_L and σ_F.

incorrect button press incurred very little penalty. In order to make the reward significant in the context of adult experience $200 was chosen as a maximum reward, of which $100 could be lost incrementally due to performance decrements. To enhance the realism, and therefore a sense of risk and consequence, participants completed all tasks in an immersive 6-degree of freedom ride motion simulator (Fig. 1A).

4.3   Reliability

In order to develop sensitive measures of TiA it is necessary to encourage a variation of TiA levels both within and across conditions. To this end we implemented two different levels of driving performance reliability; high and low. Reliability characteristics were realized by using lane and speed offsets approximately described by normal distributions with parameters specific to reliability condition as shown in Fig. 1B. The high reliability condition had narrow lane and range offset distributions whereas the distributions in the low reliability automation were broader. The low reliability automation offsets reduced the appearance of consistency over time; here, the offsets were large enough to make it appear that the automation 'wandered' gradually across the lane and following range to varying degrees based on condition.

4.4   Workload

Management of task loading was an important design constraint because of the known interaction between TIA and subjective workload; especially as affecting psychophysiological measures which are of ultimate interest in subsequent analyses. In two "full control" conditions (heading + speed control with low and high reliability) there was an inherent difference in workload owing to reduction in tasking for the human when the automation was performing well. Thus, we expected subjective workload in the high reliability, full (FH) condition to be less than in the low reliability full (FL) condition. The "speed only" conditions were introduced to allow balancing of workload across these conditions. During the speed only, high reliability (SH) condition, it was thought that subjects would primarily need to respond to lateral perturbations because the automation was near perfect in responding to longitudinal perturbations. Conversely, in speed only, low reliability (SL) conditions it would have been necessary to respond to almost all of the perturbations. To balance this circumstance across the speed only conditions, lateral perturbations were introduced more frequently in the SH as compared with the SL, thus aiming to maintain comparable overall workload in both and, importantly, allowing for inferences regarding TIA that were not confounded by effects of increased workload.

4.5   Experimental Design

The average drive time around the course for each condition lasted approximately 12 min. The two different automation capabilities were full control, i.e., both lane and range conforming ability, the second only controlled the speed of the vehicle. Automation reliability groups were high and low reliability. A 2×2 design was realized through automation type (S, F), and automation reliability (L, H); the manual run was treated as a baseline condition. The experiment consisted of five conditions; manual driving only, full automation with high reliability (FH), full automation with low reliability (FL), speed automation with high reliability (SH), and speed automation with low reliability (SL).

Psychophysiological sensors (electroencephalography (EEG), electrocardiography, galvanic skin response (GSR), and eye tracking) were fitted to each participant and then they completed a 10 min training session where they experienced both types of automation and some of the experimental tasks. After training, data collection began with onset of a manual condition, followed by the other four conditions in a counterbalanced sequence. The course was designed with straight as well as both gradual and sharply curved zones in order to change the likelihood that a trust based decision about automation use would need to be made. Surveys were administered both before the experiment and in between each condition in order to ascertain whether or not we had met our task constraints. The surveys of focus for this paper were the NASA-TLX to assess workload, and trust in automation surveys to gauge TiA levels.

5   Initial Results

Our goal was to develop a paradigm for use in studying TiA as well as neural and cognitive correlates in the operational environment of human automation systems. Previous research aimed at understanding TiA specifies particular, operationally-relevant design constraints that must be met for a successful paradigm to be developed. These include specification of a risk and consequence structure, managing the perceived reliability of automation to influence TiA, and balancing workload. An indication of successful paradigm development, therefore, would be the demonstration of having met these experimental design constraints. Here, we provide subjective survey and behavioral data indicating that our main design was effective.

TiA levels have been shown to be affected by automation reliability. Therefore, it would be expected that low reliability conditions would correspond to low TiA, whereas high reliability conditions would correspond to high TiA. Figure 2A illustrates the relationship between subjective ratings of system trustworthiness and automation reliability by condition. The TiA data were analyzed with a mixed model where reliability and type were fixed and subject data treated as random. There was a significant effect of automation type ($F(1, 71) = 3.47$, $p < 0.05$) and reliability ($F(1, 71) = 71.43$, $p < 0.01$). More important, there was also a significant interaction between automation type and reliability ($F(1, 71) = 5.0$, $p < 0.05$). Figure 2B shows that automation-related decision-making behavior, as revealed in the percentage of time the automation was engaged, reflected the change in apparent TiA as expected.

To examine whether we successfully constructed our paradigm to account for a suspected confound between subjective workload and trust in automation, we assessed the NASA-TLX. Weighted scores are shown in Fig. 3 and hypothesis tests with mixed-model ANOVA confirmed a significant automation type by reliability interaction ($F(1, 71) = 7.8$, $p < 0.05$).

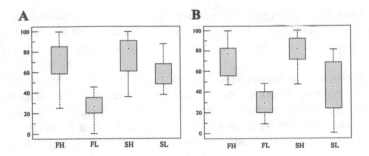

Fig. 2. (A) Subjective ratings (percent) of system trustworthiness, assessed with a visual analogue scale based on Muir (1996) and (B) percent of time automation was used when available.

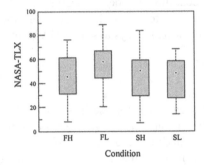

Fig. 3. Overall weighted average scores from the NASA TLX administered at the completion of each driving condition.

6   Discussion

Our aim was to develop an experimental paradigm that allows the study of TiA in the context of interactions with driving automation, an increasingly common operational environment. Behavioral and survey results indicate that we met the required design constraints derived from the TiA literature. For example, TiA levels had a clear relationship with automation reliability conditions, a key factor in TiA development. Figure 2A may also highlight the importance of predictability in TiA preservation; while SL was less trusted than SH, the SL condition appeared to be more trustworthy than the FL condition. This finding likely speaks to the issue of intersecting risk, trust, and predictability. That is, the speed control was likely experienced as generally lower risk than full control because it did not have the capability of steering into the path of an oncoming vehicle. Moreover, its following ability was so *consistently* poor in the SL condition that subjects almost always took over control immediately upon experiencing a longitudinal perturbation. Time spent using the automation should reflect TiA levels. Figure 2B shows the distribution of the percentage of time the automation mode was engaged per condition. One important variable, workload, needed to be controlled for across the speed only conditions. This was done by increasing the number of lateral perturbations that were introduced during the SH condition. Figure 3 gives NASA-TLX scores indicating that the changes made to the perturbation ratio successfully balanced subjective workload across the speed only conditions.

The behavioral and survey data indicate that our paradigm successfully achieved our goals. More importantly, in achieving the overt objectives of the study, this paradigm provides a valid start to future analysis beginning with the baseline understanding that the data were collected in accord with key constraints required for operational relevance. If our initial high level results indicated that, for instance, workload was not controlled adequately, any subsequently observed significant differences in psychophysiological variables could not be clearly attributed to TiA alone. However, more than understanding the changes in the psychophysiological variables associated with dynamic levels of TiA, is the inquiry into how changes in these variables might reflect the psychophysiological underpinning for the observed behavior, i.e., the interactions with automations, and the development of TiA. These interaction behaviors result from decisions made against a background of current psychological state which has been shown to significantly affect decision making.

Operational neuroscience studies in the context of driving automation might be aimed at understanding the psychophysiological events that support these interaction decisions, such as specific EEG and GSR features. Cognitive neuroscience research into decision making has discovered some of the neural dynamics involved in decision making. For example, fMRI studies have shown that the amygdala and the ventral striatum act to assess the valence of stimuli and that these signals are compared in the intraparietal region [17]. While operational studies necessarily use EEG rather than fMRI, these findings provide a basis for hypotheses about the cortical sources, which can be identified through localization algorithms. EEG studies aimed at understanding the cortical dynamics of complex real world decisions have identified specific frequency changes over the medial frontal regions [18]. Accompanying these neural correlates of decision making are changes in peripheral physiological and eye movement behavior. In particular, during difficult decisions, average tonic GSR magnitude increases more than if the decision was easy [19]. Eye movement, specifically gaze fixation behavior has also been associated with the cognitive processing of stimuli prior to a decision [20]. Clearly, results from cognitive neuroscience studies of decision making are fertile ground from which to generate hypotheses for further analyses in well-conducted operational experiments. We believe that our paradigm provides a research environment capable of addressing such questions.

7 Conclusion

Poor human automation integration due to mis-calibrated levels of TiA motivated an attempt to create an experimental paradigm suited to measure TiA in an operational context so that it is applicable to the real world. Because trust is a psychological state, we considered that the application of cognitive neuroscience methods rooted in psychophysiology, would be an appropriate approach to developing indicators of TiA. Typically, these methods are not used for operational neuroscience and therefore a new experimental paradigm was required. We determined through literature review what constraints were needed for successful paradigm development. We found that if indicators of TiA were to be developed that (1) there needed to be a sense of risk or consequence, (2) that there needed to be different reliabilities of the presented automations in order to manipulate TiA levels, and (3) that workload needed to be balanced across conditions. Driving was considered to provide an optimal operational environment for our research because most adults experience driving regularly and therefore would require little training. In particular, as driving automations are becoming more common driver TiA is critical, and in the driving environment, subjects would naturally interact with automations they are familiar with. Our initial results suggest that we met these goals and that our experimental paradigm provides a valid method of studying TiA and human automation interaction in an operational setting.

Acknowledgement. This research was supported by the Office of the Secretary of Defense Autonomy Research Pilot Initiative program MIPR DWAM31168, and in part by an appointment to the U.S. Army Research Postdoctoral Fellowship Program administered by the Oak Ridge Associated Universities through a cooperative agreement with the U.S. Army Research Laboratory. Research was sponsored by the Army Research Laboratory and was accomplished under Cooperative Agreement Number W911-NF-12-2-0019. The views and conclusions contained in this document are those of the authors and should not be interpreted as representing the official policies, either expressed or implied, of the Army Research Laboratory or the U.S. Government. The U.S. Government is authorized to reproduce and distribute reprints for Government purposes notwithstanding any copyright notation herein. We would also like to thank Dr. Justin Brooks and Dr. Javier Garcia for their advice.

References

1. Muir, B.M.: Trust in automation: Part I. Theoretical issues in the study of trust and human intervention in automated systems. Ergonomics **37**, 1905–1922 (1994)
2. Muir, B.M.: Operators' trust in and percentage of time spent using the automatic controllers in a supervisory process control task. Doctoral, University of Tornonto (1989)
3. Lee, J., Moray, N.: Trust, control strategies and allocation of function in human-machine systems. Ergonomics **35**, 1243–1270 (1992)
4. Prinzel, L.J., Freeman, F.G., Scerbo, M.W., Mikulka, P.J., Pope, A.T.: A closed-loop system for examining psychophysiological measures for adaptive task allocation. Int. J. Aviat. Psychol. **10**, 393–410 (2000)
5. Scerbo, M.: Adaptive automation. In: Neuroergonomics: The Brain at Work, pp. 239–252 (2006)
6. Borum, R.: The science of interpersonal trust (2010). Corritore, L., Kracher, B., Wiedenbeck, S.: On-line trust: concepts, evolving themes, a model. Int. J. Hum.-Comput. Stud. **58**, 737–758 (2003)
7. Lee, J.D., Moray, N.: Trust, self-confidence, and operators' adaptation automation. Int. J. Hum.-Comput. Stud. **40**, 153–184 (1994)
8. Lee, J.D., See, K.A.: Trust in automation: designing for appropriate reliance. Hum. Factors: J. Hum. Factors Ergon. Soc. **46**, 50–80 (2004)
9. Muir, B.M., Moray, N.: Trust in automation. Part II. Experimental studies of trust and human intervention in a process control simulation. Ergonomics **39**, 31 (1996)

10. Merritt, S.M., Ilgen, D.R.: Not all trust is created equal: dispositional and history-based trust in human-automation interactions. Hum. Factors: J. Hum. Factors Ergon. Soc. **50**, 194–210 (2008)

11. Wickens, C.D., Dixon, S.R.: The benefits of imperfect diagnostic automation: a synthesis of the literature. Theor. Issues Ergon. Sci. **8**, 201–212 (2007)

12. Parasuraman, R., Sheridan, T.B., Wickens, C.D.: A model for types and levels of human interaction with automation. IEEE Trans. Syst. Man Cybern. **30**, 10 (2000)

13. Dzindolet, M.T., Pierce, L.G., Beck, H.P., Dawe, L.A.: Misuse and disuse of automated aids. In: Proceedings of the Human Factors and Ergonomics Society Annual Meeting, p. 339 (1999)

14. Wickens, C.D.: Imperfect and unreliable automation and its implications for attention allocation, information access and situation awareness (2000)

15. Cummings, M.L., Mastracchio, C., Thornburg, K.M., Mkrtchyan, A.: Boredom and distraction in multiple unmanned vehicle supervisory control. Interact. Comput. **25**, 34–47 (2013)

16. Basten, U., Biele, G., Heekeren, H.R., Fiebach, C.J.: How the brain integrates costs and benefits during decision making. Proc. Natl. Acad. Sci. **107**, 21767–21772 (2010)

17. Davis, C.E., Hauf, J.D., Wu, D.Q., Everhart, D.E.: Brain function with complex decision making using electroencephalography. Int. J. Psychophysiol. **79**, 175–183 (2011)

18. Zhou, J., Sun, J., Chen, F., Wang, Y., Taib, R., Khawaji, A., et al.: Measurable decision making with GSR and pupillary analysis for intelligent user interface. ACM Trans. Comput. Hum. Interact. (ToCHI) **21**, 33 (2015)

19. Glaholt, M.G., Reingold, E.M.: Eye movement monitoring as a process tracing methodology in decision making research. J. Neurosci. Psychol. Econ. **4**, 125 (2011)

20. Gidlöf, K. et al.: Using eye tracking to trace a cognitive process: Gaze behaviour during decision making in a natural environment. J. Eye Mov. Res. **6**(1), 1–14 (2013)

Performance-Based Eye-Tracking Analysis
in a Dynamic Monitoring Task

Wei Du[1] and Jung Hyup Kim[2(✉)]

[1] Service Standards and Development Department,
Air China Cargo Co. Ltd., Beijing, China
uwei@airchinacargo.com
[2] Department of Industrial and Manufacturing Systems Engineering,
University of Missouri, Columbia, MO, USA
kijung@missouri.edu

Abstract. The goal of this study is to explore how ocular behavior is different in groups that possessed varying levels of performance in dynamic control tasks with complex visual components. Twenty two university students participated in this study by operating a human-in-the-loop (HITL) simulator. The participants were asked to identify unknown air track(s) and take proper actions to defend a battleship. During the experiment, a head-mounted eye-tracking device was used continuously to record participants' visual attention span regarding the normalized coordinates of their gaze points. In the current study, fixation duration was the main eye-tracking metrics. Air track identification accuracy and the NASA Task Load Index (NASA-TLX) were also used to measure participants' task performance and overall subjective mental workload.

Keywords: Mental workload · Cognitive modeling · Eye tracking analysis · Task performance · Human-in-the-loop simulation

1 Introduction

Nowadays, many complex tasks, including tasks related to monitoring the refinery process, as well as military command and control tasks, involve the use of visual information, such as display gauges and computer monitors. To carry out the responsibilities of these tasks, individuals need to conduct efficient visual searches and locate the relevant information quickly and precisely. Due to the characteristics of visual searching tasks, knowing individuals' visual attention span and their cognitive abilities is essential in evaluating task performance and display interface usability. Thus, the eye movement has become the focus of more and more studies in the last few decades. The eye-tracking technology has been successfully applied in many research areas to examine human visual attention and physiological change. The main reason that the eye-tracking method has been so widely adopted is that it provides evidence on human information processes in several aspects by using different measurements, such as fixations, saccades, scan path, pupil diameters, and so on. In our study, an experiment with the eye-tracking technology was conducted to collect the participants' eye-tracking data including eye fixation (when and where a person is looking at) and

© Springer International Publishing Switzerland 2016
D.D. Schmorrow and C.M. Fidopiastis (Eds.): AC 2016, Part II, LNAI 9744, pp. 168–177, 2016.
DOI: 10.1007/978-3-319-39952-2_17

eye fixation transition (the sequence in which their eyes are shifting from one location to another). The purpose of this study was to advance our understanding of eye movement and to explore the feasibility of various eye-tracking metrics in the visual searching task to identify differences in eye-tracking data across different performance-based groups. These quantifiable differences may be used in several ways. First, they help to further our understanding of the participants' problem-solving behaviors. Second, ocular behaviors may yield some insights into the perceptual process patterns of the participants with a high-performance level. In other words, using eye-tracking devices to identify the differences in perceptual process patterns would be advantageous in developing the appropriate manner to solve problems in some cases. Besides, the level of the participants' subjective mental workload during the task procedure was measured by using the NASA-TLX questionnaire as well. The workload result was used to develop the relationship between the participants' performance and the mental demands placed on the participants by a task.

2   Literature Review

2.1   Eye-Tracking

Eye-tracking technologies typically collect two types of eye movement information: location (where fixations tend to be directed) and duration (how long they typically remain there) within specific areas. Although it was found that human ocular behaviors showed various patterns when inspecting visual scenes, they can still be modulated, depending on cognitive demand and the characteristics of the scene (Yarbus 1967). Therefore, an eye's gaze can be considered an unbiased indicator of the focus of visual attention and the eye-tracking method has been successfully utilized to study human behavior in a wide range of research domains, such as reading (Hyönä and Niemi 1990; Rayner et al. 2006), program comprehension (Bednarik and Tukiainen 2006; Crosby and Stelovsky 1990), arithmetic problem solving (Andrá et al. 2013; Hegarty et al. 1992), multimedia learning (Tsai et al. 2012; van Gog and Scheiter 2010), and driving (Palinko et al. 2010), human-computer interaction (Granka et al. 2004; Jacob and Karn 2003). A review of the literature on eye-tracking revealed that researchers developed a wide variety of eye tracking metrics. However, one of the most commonly reported eye-tracking metrics in previous studies is fixation. There are also several derived metrics that stem from these basic measures, such as saccade, dwell (also known as gaze), scan path, and pupillary responses, such as the changes in pupil size (Jacob and Karn 2003; Poole and Ball 2005).

2.2   Fixation

Fixations, which are widely used eye tracking matrices, are commonly defined as moments in time when the eyes stay relatively stationary, taking in or "encoding" visual information. Fixation duration is one of the common metrics, both of which can be interpreted differently depending on the characteristics of the tasks. A longer duration indicates a greater difficulty in interpreting information from the object being fixated, or it may also mean that the participant is more engaging (Just and Carpenter 1976). Also, fixation duration varies as a function of the particular task such as reading, visual search and typing (Rayner 1998). It can range from 60 to 500 ms being about 250 ms on average (Liversedge and Findlay 2000). Many previous eye-tracking studies employed fixation-derived metrics and provided interesting insights into the problem-solving process. Kun et al. (Guo et al. 2006) compared the cognitive processes for inspecting several visual scenes with different characteristics and found that the face and natural images attracted similar numbers of fixations while the viewing of faces was accompanied by longer fixation durations compared with natural scenes, which provided supportive evidence to the arguments that face perception is involved in a unique cognitive process compared with non-face object or scene perception. Bednarik et al. (2006) presented a study of the comprehension processes of programmers with the help of a remote eye-tracking device. A significant difference was found on the mean fixation durations over the main areas of interest, which, they believed, indicated levels of difficultness in comprehending different parts of a program. In their study, longer durations indicated more difficulties during cognitive processing. A more recent research done by Tsai et al. (2012) examined the students' visual attention when solving a multiple-choice problem. Based on the analysis, they suggested that the relevant information was fixated longer than the irrelevant one, and participants paid the most attention (longer total fixation durations) to their preferred answers before making decisions. Table 1 is a summary of the main fixation-derived metrics.

Table 1. Fixation-derived metrics with interpretations from literatures

Eye movement metric	Interpretation	Reference
Fixation duration mean	Longer durations indicate a greater difficulty in interpreting information from the object	Just and Carpenter (1976); Jacob and Karn (2003)
	Longer durations mean that the object is more engaging in some way	Just and Carpenter (1976)
Total fixation number	More overall fixations indicate more searching or less efficient search	Goldberg and Kotval (1999)
Fixation number per area of interest	A higher number of fixations indicates more importance or a more noticeable area in problem-solving tasks	Poole et al. (2005)
	A higher number of fixations means that an item is actually harder to recognize in reading tasks	Poole et al. (2005)

2.3   Workload Measurement

Currently, applied research has paid much attention to the human workload study to ensure high levels of safety, health, and comfort and the long-term productive efficiency of the operator. Eggemeier (1988) defined mental workload as "the degree of processing capacity that is expanded during task performance"; in other words, the mental workload is a specification of the amount of information processing capacity that is used for task performance. There are several factors that are said to be related to mental workloads from the both task side (such as complexity, the difficulty of the task) and the individual side (such as effort expended by the operator) (De Waard and Studiecentrum 1996). Thus, a multidimensional measurement method is needed to scale the workload accurately. A number of tools for the evaluation and prediction of workload are available such as the National Aeronautics and Space Administration – Task Load Index (Hart and Staveland 1988), the Subjective Workload Assessment Technique (Reid and Nygren 1988), and the Workload Profile (Tsang and Velazquez 1996). Among these workload assessment instruments, NASA-TLX, which is a multifaceted tool for assessing subjective workload, has been applied in many domains and is widely accepted as one of the strongest tools for reporting perceptions of workload.

3 Method

In this study, the anti-air warfare coordinator (AAWC) human-in-the-loop test bed was used (Kim et al. 2015; Macht et al. 2014). Within this interactive simulator, participants handled identifying the unknown air tracks based on the engagement rules and the air tracks' information that they gathered from the radar simulation system. The main reason for choosing this test platform was that it was a relatively complex task with dynamic visual components. Also, some studies have contributed to understanding the eye movement behaviors using this platform or other similar platforms. Thus, further evidence was needed to support the reliability of eye-tracking metrics in complex dynamic tasks such as the AAWC simulation (Fig. 1).

Fig. 1. Experimental environment

The twenty-two participants were divided into three groups based on their performance: high-accuracy group, medium-accuracy group, and low-accuracy group. Analysis of the data revealed some statistical changes in ocular behavior and mental workload among the three groups.

4   Data Collection

4.1   AAWC Simulation Task Performance

The accuracy of the identified unknown air tracks was calculated to determine the levels of the participants' AAWC simulation task performance. As discussed above, we only focused on the performance of identification task in the current study. Thus, if and only if both the primary identification (such as friendly, hostile, and etc.) and the type identification (such as strike, commercial aircraft, helicopter and etc.) of an unknown air track were identified correctly before the end of a scenario, the result would be accounted as a right identification (Kim 2014; Kim et al. 2011).

$$\text{Accuracy} = \frac{\text{Total number of correctly identified air tracks}}{\text{Total number of unknwon air tracks in the scenario}} \tag{1}$$

4.2   Area of Interest (AOI)

The AOI refers to the specified interested areas in the visual field in which the gaze point lies. Five AOIs were defined for further data analysis (See Fig. 2). They were radar screen AOI, menu bar AOI, data panel AOI, track profile AOI, and EWS AOI. Although being important components of the AAWC simulation task, some areas were defined as outside areas and were excluded from eye-tracking analysis for the following reasons. The AAWC simulator guide board was excluded due to the small amount of attention allocation. The air track behavior board area and the system response panel area were excluded due to the low eye-tracking accuracy on these two areas. Also, since the purpose of this experiment was to explore the differences on information process, the data input panel area was excluded as well as it did not contain much relative information that can be used to perceive the situation of the air space.

Fig. 2. Environment layout and areas of interest (AOIs)

4.3   Eye-Movement Data

The raw data recorded by the eye-tracking system were typically in terms of the normalized coordinates. From this raw data, fixations were calculated and derived. In this study, fixation occurred when a participant's gaze stabilized over at least 100 ms, and the visual angle degree was less than 1°. The fixation duration was calculated as the entire period of the fixation, in other words, the difference between stop and start time of the fixation.

5   Results

5.1   AAWC Simulation Task Performance

Although the participants went through the same experiment training phases, their mastering of the knowledge and skills were varied. The analysis of the participants' identification accuracy was performed to determine the level of performance. According to the identification result, twenty-two participants were divided into three groups with the different performance levels: the high-accuracy group (8 participants), the medium-accuracy group (7 participants), and the low-accuracy group (7 participants).

The followings are the criteria for grouping.

- High-Accuracy Group: The average accuracy of the two day's experiment was no less than 75 % and the accuracy on each day was no less than 70 % (This criteria filtered out about top 25 % participants with relatively high performance).
- Medium-Accuracy Group: The average accuracy of the two day's experiment was between 50 % and 65 % and none of the accuracy on each day was lower than 40 % or higher than 70 % (This criteria filtered out about medium 25 % participants with relatively medium performance).
- Low-Accuracy Group: The average accuracy of the two day's experiment was no more 45 % and the accuracy on each day was no more than 50 % (This criteria filtered out about bottom 25 % participants with relatively low performance).

5.2   Fixation Durations on AOIs

In order to determine the information processing time on the AOIs, we investigated the fixation duration time. The AAWC test platform consisted of five areas of interest (AOIs), which were radar screen AOI, menu bar AOI, data panel AOI, track profile AOI, and EWS AOI. Eye-tacking data was filtered out when eye gaze point fell outside these areas. Thus, fixation duration time was measured only for these five discrete AOIs. The results of fixation durations on each AOI are presented in Table 2.

Table 2. Mean fixation durations and standard deviations (in parentheses) over the main AOIs for high-accuracy, medium-accuracy and low-accuracy groups

Group	Radar screen	Data panel	Track profile	EWS	Menu bar
High-accuracy	0.35749 (0.38364)	0.30281 (0.25152)	0.23630 (0.13387)	0.21711 (0.11929)	0.29523 (0.20356)
Medium-accuracy	0.37653 (0.40092)	0.30621 (0.23886)	0.23693 (0.14868)	0.22769 (0.12502)	0.30204 (0.21927)
Low-accuracy	0.37107 (0.39057)	0.34800 (0.35655)	0.24653 (0.16285)	0.22862 (0.12380)	0.30735 (0.22311)

5.3   NASA-TLX

According to the assessment questionnaire, the mean of the overall weighted NASA-TLX score for the high-accuracy group was 42.77 (SD = 15.60), which was the lowest among all the three groups. The mean of the overall weighted NASA-TLX score for medium-accuracy group was of the medium level, which was 49.54 (SD = 16.55). However, as for the low-accuracy group, the overall weighted NASA-TLX score increased up to 56.59 (SD = 19.78) which was higher than any other groups. The ANOVA analysis showed that the three groups' NASA-TLX scores were significantly different ($F(2,129) = 7.24$, $p < 0.001$) (Table 3). The medium-accuracy group and the low-accuracy group's scores were significantly higher than the high-accuracy group's (all p-values <0.05). However, no significant difference was found between medium-accuracy group and low-accuracy group (p = 0.076).

Table 3. NASA-TLX results of high-accuracy, medium-accuracy and low-accuracy groups

Group	Mean NASA TLX	Standard deviation	F value	P value
High-accuracy	42.77	15.60	$F(2,129) = 7.24$	0.001
Medium-accuracy	49.54	16.55		
Low-accuracy	56.69	19.78		

6   Discussion and Conclusion

By analyzing the result, we found that the fixation duration varied systematically as a function of the five different AOIs as expected. It means that the fixation duration is linked to the human's processing time. A longer duration indicates that more effort is expended in interpreting information during a problem-solving process. We discovered that a comparison of fixation duration data produced clear differences corresponding to the known levels of the performance in our study. It indicates that the processing time to comprehend information from different AOIs varied a lot and obviously followed the same trend across all three groups. The longest mean fixation duration belonged to the radar screen AOI, and the second-longest fixated AOI was the data panel AOI followed by the menu bar AOI, while the EWS AOI and the track profile AOI demanded the lowest mean fixation durations. This can be related to the different difficulty levels for participants to encode the information from each AOI. Participants demanded the longest fixation duration on the radar screen AOI, which could be explained by the fact that this AOI displayed only graphic symbols, which indicated a lot of implicit information including current track identification, direction, speed, location, and behavior. It had a longer solution path and required more effort in order to extract the useful information from this AOI. As for the data panel AOI, which displayed the air track's parameters in digits, such as altitude and speed, the information on this AOI was much more straightforward compared with the information on the radar screen AOI. It saved participant effort in encoding implicit information from the symbols on the radar screen AOI. Additionally, the information on the EWS AOI and track profile AOI was the easiest to process due to the fact that these only displayed the direct correspondence between the track's real identification and the typical parameter/sensor code/track model, which required no further processing.

In addition, a significant difference on NASA-TLX scores was observed across the groups with different performance levels. It was found that participant performance had an obvious negative correlation with the NASA-TLX score. As the group's identification accuracy significantly increased, the NASA-TLX score decreased. This relationship revealed the fact that the high-accuracy group experienced a relatively low overall mental workload when performing the radar monitoring task. In other words, the mental demands expended during the task were lower for those with higher performances. The result can possibly be explained by the fact that if participants had a stably high performance, which means they could understand the task well, be aware of the situation correctly and interact with the radar simulator step by step in an orderly manner, they experienced less mental workload and might have felt more relax during the tasks. However, if participants had relatively worse performances, which means they experienced more difficulties when struggling in the task, they might be more stressful and expended higher mental workload. In a nutshell, the NASA-TLX could be used as a supportive source for measuring performance levels. However, there is a limitation on NASA-TLX. Though the NASA-TLX scores showed an apparent inverse pattern from the performance level, the difference in NASA-TLX scores between the medium-accuracy group and the high-accuracy group was not significant. It means that this subjective response might not precisely correspond to the participants' experience. In addition, NASA-TLX is a post-session subjective assessment. Hence, it could not account for rapid changes in mental workload.

In this research, we used the anti-air warfare coordinator (AAWC) radar simulator as the experiment test bed to explore how ocular behavior is different in high-accuracy, medium-accuracy, and low-accuracy groups. The findings suggested that eye tracking data may potentially be a reliable source to identify differences in problem-solving behaviors among performance-based groups in several aspects, and may provide insights into the cognitive process to interpret participant performance further.

References

Andrá, C., Lindström, P., Arzarello, F., Holmqvist, K., Robutti, O., Sabena, C.: Reading mathematics representations: an eye-tracking study. Int. J. Sci. Math. Educ. **13**(2), 237–259 (2013)

Bednarik, R., Tukiainen, M.: An eye-tracking methodology for characterizing program comprehension processes. Paper Presented at the Proceedings of the 2006 Symposium on Eye Tracking Research and Applications (2006)

Crosby, M.E., Stelovsky, J.: How do we read algorithms? A case study. Computer **23**(1), 25–35 (1990)

De Waard, D., Studiecentrum, V.: The measurement of drivers' mental workload. Groningen University, Traffic Research Center Netherlands (1996)

Eggemeier, F.T.: Properties of workload assessment techniques. Adv. Psychol. **52**, 41–62 (1988)

Granka, L.A., Joachims, T., Gay, G.: Eye-tracking analysis of user behavior in WWW search. Paper Presented at the Proceedings of the 27th Annual International ACM SIGIR Conference on Research and Development in Information Retrieval (2004)

Goldberg, J.H., Kotval, X.P.: Computer interface evaluation using eye movements: methods and constructs. Int. J. Ind. Ergonomics, **24**(6), 631–645 (1999)

Guo, K., Mahmoodi, S., Robertson, R.G., Young, M.P.: Longer fixation duration while viewing face images. Exp. Brain Res. **171**(1), 91–98 (2006)

Hart, S.G., Staveland, L.E.: Development of NASA-TLX (task load index): results of empirical and theoretical research. Adv. Psychol. **52**, 139–183 (1988)

Hegarty, M., Mayer, R.E., Green, C.E.: Comprehension of arithmetic word problems: Evidence from students' eye fixations. J. Educ. Psychol. **84**(1), 76 (1992)

Hyönä, J., Niemi, P.: Eye movements during repeated reading of a text. Acta Psychol. **73**(3), 259–280 (1990)

Jacob, R., Karn, K.S.: Eye tracking in human-computer interaction and usability research: Ready to deliver the promises. Mind **2**(3), 4 (2003)

Just, M.A., Carpenter, P.A.: Eye fixations and cognitive processes. Cogn. Psychol. **8**(4), 441–480 (1976)

Kim, J.H.: Simulation training in self-regulated learning: investigating the effects of dual feedback on dynamic decision-making tasks. In: Zaphiris, P., Ioannou, A. (eds.) LCT 2014, Part I. LNCS, vol. 8523, pp. 419–428. Springer, Heidelberg (2014)

Kim, J.H., Chan, T., Du, W.: The learning effect of augmented reality training in a computer-based simulation environment. In: Zaphiris, P., Ioannou, A. (eds.) LCT 2015. LNCS, vol. 9192, pp. 406–414. Springer, Heidelberg (2015)

Kim, J.H., Rothrock, L., Tharanathan, A., Thiruvengada, H.: Investigating the effects of metacognition in dynamic control tasks. In: Jacko, J.A. (ed.) Human-Computer Interaction, Part I, HCII 2011. LNCS, vol. 6761, pp. 378–387. Springer, Heidelberg (2011)

Liversedge, S.P., Findlay, J.M.: Saccadic eye movements and cognition. Trends Cogn. Sci. **4**(1), 6–14 (2000)

Macht, G.A., Nembhard, D.A., Kim, J.H., Rothrock, L.: Structural models of extraversion, communication, and team performance. Int. J. Ind. Ergon. **44**(1), 82–91 (2014)

Palinko, O., Kun, A.L., Shyrokov, A., Heeman, P.: Estimating cognitive load using remote eye tracking in a driving simulator. Paper Presented at the Proceedings of the 2010 Symposium on Eye-Tracking Research and Applications (2010)

Poole, A., Ball, L.J.: Eye tracking in human-computer interaction and usability research: current status and future. Prospects. In: Ghaoui, C. (ed.) Encyclopedia of Human-Computer Interaction. Idea Group. Inc, Citeseer, Pennsylvania (2005) level of the additional computational effort. The capability of the method in [8] is evaluated in the presence of Gaussian noise. The probability density function of the noise (PDF) affecting the CT images can also be considered as Gaussian [3,4], therefore the method is suitable for denoising of CT images.

In general, to perform the evaluation of the denoising methods classical measures as Peak Signal to Noise Ratio (PSNR) or Structural Similarity (SSIM) index [9] are used. These measures work with the denoised image and the original noiseless image as the reference. Unfortunately, such reference is not available in case of real CT images, thus the PSNR and SSIM cannot be used. For the purpose of the quality evaluation of the collaborative filtering method we have established our own evaluation process consisting of measurement and visual inspection of the resulting denoised images. The measurement is based on the assumption, that if we divide the CT image into appropriate number of significant segments that correspond to individual body tissues, each segment should have as low variance (standard deviation) in its intensity as possible. Division into image segments was done by k-means segmentation algorithm [7] with image pre-filtering by averaging filter of size 8×8 pixels. Number of segmented areas was set to 12. To obtain only one evaluating value we have used weighted average of standard deviations of all segmented areas. As weights we have used the number of pixels present in each of the segmented areas. In the visual inspection process we use the assumption, that the denoising method should remove only the noise without distorting important details in the image. If we therefore visualize the difference between the original noisy image and the denoised image, only the noise structure should be visible without any contours or other recognized features of the image objects.

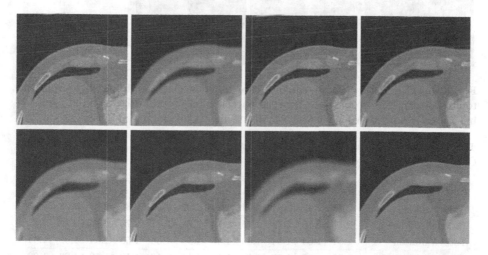

Fig. 4. Image detail of different denoising methods, from left to right and top to bottom; original CT image, Anisotropic filter, Wavelet packets filter, Total Variation filter, Gaussian blur filter, Adaptive Wiener filter, NL Means filter, Collaborative filtering

134   P. Strakos et al.

We have used several well known and often used denoising methods based on the survey published in [10] to compare with the collaborative filtering method. Namely we have used Anisotropic diffusion [11], Wavelet packets [12], Total Variation denoising [13,14], Gaussian blur [15], Adaptive Wiener filter [16], Non-Local Means filter [17] and the elaborated collaborative filtering method. The optimal parameters for each of the denoising algorithms were determined empirically.

Denoising capabilities were tested on series of 10 consecutive images. Results of the comparison can be seen in Figs. 4, 5 and Table 2. The results prove the strong denoising capability of the collaborative filtering method and show its prevailing quality over all other tested methods. The only drawback of the method is its extensive computational burden. If one considers the application of the method for denoising of large series of CT images, the total runtimes can go to hours. Especially for this reason the effort was taken and we provide

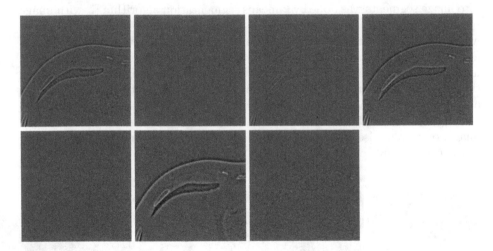

Fig. 5. Image detail with noise eliminated by different methods, from left to right and top to bottom; Anisotropic filter, Wavelet packets filter, Total Variation filter, Gaussian blur filter, Adaptive Wiener filter, NL Means filter, Collaborative filtering

Table 2. Estimated weighted average standard deviation $\bar{\sigma}_{w1} \div \bar{\sigma}_{w10}$ for 10 consecutive CT images; $\bar{\bar{\sigma}}_w$ mean value of 10 measurements

Filter	$\bar{\sigma}_{w1}$	$\bar{\sigma}_{w2}$	$\bar{\sigma}_{w3}$	$\bar{\sigma}_{w4}$	$\bar{\sigma}_{w5}$	$\bar{\sigma}_{w6}$	$\bar{\sigma}_{w7}$	$\bar{\sigma}_{w8}$	$\bar{\sigma}_{w9}$	$\bar{\sigma}_{w10}$	$\bar{\bar{\sigma}}_w$
Anisotropic	2.79	2.82	2.84	2.94	2.99	2.92	2.95	2.85	2.83	2.78	2.87
Wavelet packets	2.77	2.79	2.78	2.78	2.79	2.78	2.77	2.76	2.76	2.75	2.77
Total Variation	1.54	1.54	1.54	1.53	1.56	1.56	1.55	1.55	1.54	1.51	1.54
Gaussian blur	1.05	1.07	1.07	1.08	1.11	1.11	1.10	1.10	1.10	1.03	1.08
Adaptive Wiener	1.37	1.38	1.38	1.38	1.40	1.41	1.40	1.40	1.39	1.34	1.38
NL Means	1.46	1.47	1.50	1.48	1.52	1.54	1.50	1.52	1.53	1.52	1.50
Collaborative filtering	**0.98**	**1.00**	**1.01**	**1.02**	**1.06**	**1.07**	**1.06**	**1.05**	**1.06**	**1.02**	**1.03**

parallel implementation of the method. Process of parallelization and the typical scalability tests of the algorithm are presented in the following section.

4 Parallel Implementation

Parallelization of the filtering algorithm has been programmed in Matlab R2015a, which serves as a rapid prototyping tool for evaluation of the performed parallelization concept of the algorithm. For users aiming at parallelization of their codes, Matlab offers Parallel Computing Toolbox [18]. Implementation of the toolbox is based on MPI routines, therefore multiple compute nodes can be utilized [19] and distributed memory is used. Toolbox is not limited to specific MPI library. Users can choose their own library. For example Intel MPI supporting the fast InfiniBand communication standard can be used [20]. Parallelization concept in Matlab uses term job for the large operation that needs to be run. Typically it is the program you aim to parallelize. The job is broken down into tasks. Dimension of each task depends on the user. Since tasks can run simultaneously on the cluster, the parts of code that are assigned to individual tasks are in general those parts of program that can be parallelized. Tasks can be independent or they can communicate between each other. The Matlab session in which you define the job and its tasks is called the client session. The Matlab session performing the task computations is the worker session. The scheduler distributes the job's tasks to available workers, after submitting the job. Each worker is evaluating exactly one task in one time instant. By default, Matlab runs one worker per core.

Parallel version of the filtering algorithm has been created by modifying its sequential implementation described in Sect. 2 and Fig. 1. Parallel version in form of pseudo-code is described below. The most extensive part of the code is grouping by matching similar image blocks within predefined smaller area of the image. For this purpose we distribute the whole image to each core (worker) together with the index vector corresponding to image pixels of reference blocks specific to smaller areas of the image where grouping and matching is performed. This whole operation is performed repeatedly in different areas of the image, but covering the whole image. It is done during creation of the initial image estimate and similarly also during the creation of the final image estimate, see Fig. 2. In pseudo-code this is represented by the rows 3.(b)i. and 5.(b)i. Together with other operations that are performed repetitively 3.(b)ii., 3.(b)iii. and 5.(b)ii., 5.(b)iii., the core of the algorithm is covered. All repetitive actions are enclosed in for loops. There are two main for loops in the algorithm, 3.(b)–(c) and 5.(b)–(c). Those loops cannot run simultaneously, because all of the iterations of the loop 3.(b)–(c) have to be finished before aggregation in point 4., which has to be finished before the loop 5.(b)–(c) can start. Our effort to parallelize the code thus focused on parallelizing the two main for loops subsequently. We have split each loop to exactly $(N_{cores} - 1)$ smaller loops, where N_{cores} is the number of assigned cores to the job. Smaller loops of the first and then the second main loop were assigned to individual tasks and each task was dedicated to exactly one core and one worker. Although we reserve N_{cores} for the job one core was always reserved for data distribution, sending and collecting of the results and to client session. We have used Matlab setting for creating communicating tasks, even though each task ran independently without exchanging data with the other simultaneously running tasks. In this way we could synchronize the tasks and provide the effective combination of serial and parallel parts of the job within Matlab.

1. Load image
2. Set parameters and create transformation matrices for Step 1, Step 2
3. Step 1 - Calculate the group estimates
   (a) Set positions of reference blocks
   (b) **for** $i = 1 \div M_1$ number of reference blocks in the image
      i. Do grouping by matching
      ii. Realize the collaborative filtering
      iii. Compute the weight value for the composed group of blocks
   (c) **end**
4. Step 1 - Aggregate the group estimates, weights and compute the initial image estimate
5. Step 2 - Calculate the group estimates
   (a) Set positions of reference blocks
   (b) **for** $i = 1 \div M_2$ number of reference blocks in the image
      i. Do grouping by matching in the initial image estimate
      ii. Collect image blocks from noisy image
      iii. Compute the Wiener coefficients
      iv. Compute the weight value corresponding to the filtered group of blocks
      v. Realize the collaborative Wiener filtering
   (c) **end**
6. Step 2 - Aggregate the group estimates, weights and compute the final image estimate

5   Results

For all tests concerning the parallel implementation and its comparison with the sequential implementation up to 4 regular compute nodes of Anselm supercomputer have been used. Every node is equipped with 2x Intel Sandy Bridge E5-2470 2.3 GHz CPUs [21]. Each CPU has 8 cores, thus up to 16 cores per node were available. In total 64 cores have been used. This particular hardware configuration was chosen to test the scaling capabilities of the proposed algorithm beyond the single node. From total sum of 64 cores one core was always reserved for data distribution and collection of the partial results as a client. Rest of the cores were involved as workers in calculation of the tasks. So minimally one client and one worker occupying two cores and maximally one client and 63 workers occupying 64 cores were used. The provided comparisons are evaluated for 1

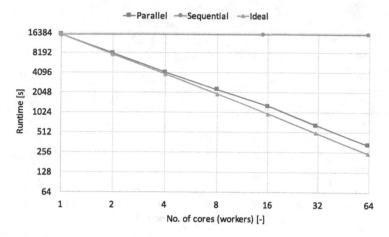

Fig. 6. Results of the strong scalability; comparison between ideal, performed parallel and sequential implementation (Color figure online)

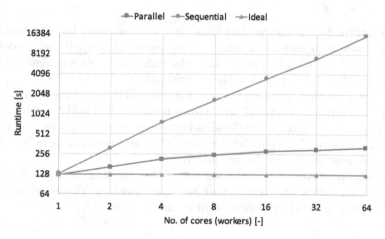

Fig. 7. Results of the weak scalability; comparison between ideal, performed parallel and sequential implementation (Color figure online)

to 63 cores of workers which is comparable to the sequential implementation running on 1 to 63 cores.

In Fig. 6 results of strong scalability of the performed parallelization are shown. Size of the problem is kept constant (CT image of typical size 512×512 pixels has been used) while number of cores is being increased. Logarithmic scales have been used on both axes to better show the difference between ideally parallel and performed parallel version of the algorithm.

In Fig. 7 results of weak scalability can be seen. In this test core load is kept constant, meaning that for each increase in number of used cores (workers) size of the whole problem is equivalently increased too. In terminology of Parallel

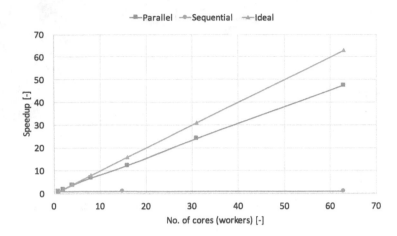

Fig. 8. Algorithm speed-up; comparison between ideal parallel, performed parallel and sequential implementation (Color figure online)

Computing Toolbox it means that task sizes are constant, but the number of tasks increases with number of added cores (workers). Logarithmic scales have been used on both axes to better show the difference between ideally parallel and performed parallel version of the algorithm.

In Fig. 8 algorithm speed-up by parallelization is shown. Size of the problem in this test is kept constant (CT image of typical size 512×512 pixels has been used) while number of cores (workers) is being increased. With every added core algorithm speed-up of the parallel versions (ideal and performed) in comparison to the sequential version are being shown.

Comparison between sequential version of the algorithm and parallel version using only one client and one worker was also studied. The overhead of the parallel version using client-server architecture is about 5 %.

6   Conclusion

The aim of the paper was twofold. First evaluation of the collaborative filtering method in terms of denoising quality of medical CT images was provided. Superiority of the method over all the other evaluated methods in term of the superior denoising quality has been shown. It should be pointed out that objective of this paper is not to compare different denoising methods in terms of time necessary for denoising. Direct comparison of denoising times of the different methods could not be objectively shown either, since some of the algorithms used in the comparison are in Matlab MEX functions (C code).

Second aim of the paper focused on parallelization of the collaborative filtering method. The filtering method is computationally extensive and its parallelization showed possible speed-up in comparison to regular sequential version of the algorithm. We have used Matlab to test the parallelization of the code. Matlab is ideal rapid prototyping tool for proving the concept of the parallelization.

Final solution for practical usage of the method would have to be implemented in C or C++, but this was not the objective. We also have not studied possible influence on the denoising time of the parallelized algorithm if we alter the parameters of the method, because the specific values used in this paper proved themselves to be the optimal ones.

Concerning the results of the proposed parallelization it speeds-up the filtering algorithm dramatically. Due to the MPI concept used behind the Matlab's Parallel Computing Toolbox the proposed parallelization can also scale over more computing nodes. The performed parallelization of the collaborative filtering method is therefore capable to fulfil high demands on denoising quality in the area of medical imaging together with efficient speed-up of the algorithm.

Acknowledgements. This paper has been elaborated in the framework of the project New Creative Teams in Priorities of Scientific Research, reg. no. CZ.1.07/2.3.00/30.0055, supported by Operational Programme Education for Competitiveness and co-financed by the European Social Fund and the state budget of the Czech Republic. The work was also supported by the European Regional Development Fund in the IT4Innovations Centre of Excellence project (CZ.1.05/1.1.00/02.0070) and the Project of Major Infrastructures for Research, Development and Innovation of Ministry of Education, Youth and Sports with reg. num. LM2011033. Authors acknowledge the support of VSB-TU Ostrava under the grant SGS SP2015/189.

References

1. Hsieh, J.: Computed Tomography: Principles, Design, Artifacts, and Recent Advances. SPIE, Bellingham (2002)
2. Kak, A.C., Slaney, M.: Principles of Computerized Tomographic Imaging. SIAM, Philadelphia (2001)
3. Lu, H., Li, X., Hsiao, I.T., Liang, Z.: Analytical noise treatment for low-dose CT projection data by penalized weighted least-squares smoothing in the K-L domain. In: Proceedings of SPIE. Medical Imaging, pp. 146–152 (2002)
4. Lei, T., Sewchand, W.: Statistical approach to X-ray CT imaging and its applications in image analysis part I: statistical analysis of X-ray CT imaging. IEEE Trans. Med. Imaging **11**, 62–69 (1992)
5. Dabov, K., Foi, A., Katkovnik, V., Egiazarian, K.: Image denoising by sparse 3D transform-domain collaborative filtering. IEEE Trans. Image Process. **16**(8), 2080–2095 (2007)
6. Lebrun, M.: An analysis and implementation of the BM3D image denoising method. Image Process. Line **2**, 175–213 (2012)
7. Strakos, P., Jaros, M., Karasek, T., Riha, L., Jarosova, M., Kozubek, T., Vavra, P., Jonszta, T.: Parallelization of the image segmentation algorithm for Intel Xeon Phi with application in medical imaging. In: 4th International Conference on Parallel, Distributed, Grid and Cloud Computing for Engineering. Civil-Comp Press, Stirlingshire (2015)
8. Katkovnik, V., Foi, A., Egiazarian, K., Astola, J.: From local kernel to nonlocal multiple-model image denoising. Int. J. Comput. Vis. **86**(1), 1–32 (2010)

9. Wang, Z., Bovik, A.C., Sheikh, H.R., Simoncelli, E.P.: Image quality assessment: from error visibility to structural similarity. IEEE Trans. Image Process. **13**(4), 600–612 (2004)
10. Buades, A., Coll, B., Morel, J.M.: On image denoising methods. SIAM Multiscale Model. Simul. **4**(2), 490–530 (2005)
11. Perona, P., Malik, J.: Scale-space and edge detection using anisotropic diffusion. IEEE Trans. Pattern Anal. Mach. Intell. **12**(7), 629–639 (1990)
12. Antoniadis, A., Oppenheim, G.: Wavelets and Statistics. Lecture Notes in Statistics, vol. 103. Springer, Heidelberg (1995)
13. Rudin, L.I., Osher, S., Fatemi, E.: Nonlinear total variation based noise removal algorithms. Phys. D: Nonlinear Phenom. **60**(1–4), 259–268 (1992)
14. Beck, A., Teboulle, M.: Fast gradient-based algorithms for constrained total variation image denoising and deblurring problems. IEEE Trans. Image Process. **18**(11), 2419–2434 (2009)
15. Sonka, M., Hlavac, V., Boyle, R.: Image Processing, Analysis and Machine Vision. Thomson, Toronto (2006)
16. Lim, J.S.: Two-Dimensional Signal and Image Processing. Prentice Hall, Englewood Cliffs (1990). Equations 9.26, 9.27, and 9.29
17. Buades, A., Coll, B., Morel, J.M.: A non-local algorithm for image denoising. In: Proceedings of Conference on Computer Vision and Pattern Recognition, vol. 2, pp. 60–65 (2005)
18. Parallel Computing Toolbox^{TM} User's Guide. The MathWorks, Natick (2015)
19. Hager, G., Wellein, G.: Introduction to High Performance Computing for Scientists and Engineers. CRC Press, Boca Raton (2011)
20. Intel MPI Library. https://software.intel.com/en-us/intel-mpi-library/
21. Hardware Overview IT4I Docs. https://docs.it4i.cz/anselm-cluster-documentation/hardware-overview

On Modeling of Bodies with Holes

Jan Franců[(✉)]

Faculty of Mechanical Engineering, Brno University of Technology,
Brno, Czech Republic
francu@fme.vutbr.cz

Abstract. Modeling of bodies with holes leads to boundary value problems for P. D. E. on perforated domains, i. e. multiply connected domains. The contribution aims to point out the problem of boundary conditions on the holes which depend on physical background of the problem.

Two examples are discussed. Thermal conduction in a perforated body with thermally insulated holes leads to homogeneous Newmann boundary conditions on the holes. The case of torsion of a bar with perforated profile leads to different situation. The usual boundary conditions are not sufficient, an unusual integral condition must be added. These boundary conditions are derived directly using the potentiality condition. The variational formulation of the problem yields the same result. The torsion problem is illustrated by examples of solutions computed in [FR].

Keywords: Body with holes · Boundary conditions on hole · Torsion of a bar · Profile with holes · Airy stress function · Integral boundary condition

1 Introduction

Behavior of a body occupying volume Ω can be modeled by a boundary value problem for partial differential equation on the domain (i.e. open connected set) Ω. Let us denote the main unknown by u, it can be temperature, displacement, potential, stress, concentration, etc.; it can be scalar or vector function on the domain Ω. Since the equation itself does not provide unique solution, the equation on Ω must be completed by boundary conditions describing the situation on the boundary $\partial\Omega$ of the volume Ω. If the body is not simply connected, i. e. the body has some holes, then its boundary $\partial\Omega$ consists of two different parts: the outer boundary and boundaries of the holes.

The contribution aims to point out the problem of boundary conditions on the holes in case of a two dimensional perforated domain Ω with holes Ω_i and its boundaries $\Gamma_i = \partial\Omega_i$, for notation see Sect. 2. What are necessary and sufficient conditions on Γ_i? It depends on the physical background of the problem, as we shall see.

The first example, see Sect. 3, deals with thermal conduction in a volume with thermally insulating holes. This case leads to the Neumann conditions on boundaries Γ_i of the holes.

© Springer International Publishing Switzerland 2016
T. Kozubek et al. (Eds.): HPCSE 2015, LNCS 9611, pp. 141–151, 2016.
DOI: 10.1007/978-3-319-40361-8_10

The second example, see Sects. 4, 5, 6 and 7, deals with torsion of a bar with perforated profile Ω. The model leads to Poisson equation $-\Delta u = 2$ on the domain Ω. We derive the boundary value problem including an unusual additional integral condition on the holes. This condition follows from the potentiality condition for the deflection, see Sect. 5. This argumentation seems to be new. In Sect. 6 these boundary conditions are derived from the variational formulation.

Variational formulation of the problem (without introducing boundary conditions on the wholes) is studied also in [L]. In [S] the torsion problem for multiply connected domain is also studied. The analysis is based on the general plane-strain solution and its compatibility conditions. The reasoning and the obtained conditions are much more complicated, the results comprise more general problems.

The problem for a domain with holes can be reached also by a limit procedure. The holes Ω_i are considered to be "filled in" with a material characterized by material coefficient q. In case of thermally insulating holes we consider a sequence of problems with q tending to zero, see Sect. 3. In case of perfectly conducting holes the sequence of problems with q tending to infinity can be considered. The contribution is closed by examples. The exact solutions for the ring and a "broken ring" profiles from [FR] are shown and the results are compared.

2   Notation and Formulation of the Problem

Although similar 3D problems can be studied we shall restrict to 2D problems and adopt the following notation and assumptions on sets and boundaries:

- Ω_0 is a bounded open simply connected set in the plane $\mathbb{R}^2$ representing the body including the volume occupied by the holes.
- $\Omega_1, \ldots, \Omega_k$ is a finite family of simply connected disjoint open subsets of Ω_0 representing the "holes" in the body.
- Ω is the space occupied by the material. Clearly $\Omega = \Omega_0 \setminus (\overline{\Omega}_1 \cup \overline{\Omega}_2 \cup \cdots \cup \overline{\Omega}_k)$. It is a multiply connected open bounded set.
- $\Gamma_0 = \partial\Omega_0$ is the outer boundary of the volume Ω.
- $\Gamma_i = \partial\Omega_i$, $i = 1, \ldots, k$ is boundary of the i-th hole Ω_i. All the boundaries are mutually disjoint and $\partial\Omega = \Gamma_0 \cup \Gamma_1 \cup \cdots \cup \Gamma_k$.
- Each Γ_i is a simple closed piecewise smooth curve. Γ_0 is oriented counterclockwise, while curves $\Gamma_1, \ldots, \Gamma_k$ are oriented clockwise, see Fig. 1.
- $n = (n_x, n_y)$ is the unit normal vector and $t = (t_x, t_y)$ the unit tangent vector to $\partial\Omega$. Since the boundaries are piecewise smooth, the vectors n and t exist on each Γ_i except for isolated points. The tangent and normal vectors are connected by the relation $t_x = -n_y$ and $t_y = n_x$.

Let us point out that the outer normal vector n on Γ_i, $i \geq 1$ is directed inside the hole Ω_i since it is outer with respect to Ω, see Fig. 1. We shall deal with the following equation

$$- \operatorname{div}(a \cdot \nabla u) \equiv -\frac{\partial}{\partial x}\left(a \cdot \frac{\partial u}{\partial x}\right) - \frac{\partial}{\partial y}\left(a \cdot \frac{\partial u}{\partial y}\right) = f \qquad \text{on } \Omega \ , \qquad (1)$$

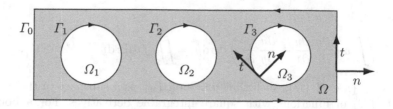

Fig. 1. Body with holes, orientation of tangent and normal vectors.

where a is a given positive constant (or function $a(x, y)$ on Ω) describing property of the material and f a given function describing the volume impacts. On the outer boundary Γ_0 the equation is completed by boundary conditions. It can be either the Dirichlet boundary condition prescribing value u

$$u = \overline{u} \qquad \text{on } \Gamma_0 , \tag{2}$$

or the Neumann boundary condition prescribing the normal flux $g = -a \frac{\partial u}{\partial n}$

$$g \equiv -a \frac{\partial u}{\partial n} = \overline{g} \qquad \text{on } \Gamma_0 . \tag{3}$$

In case of the so-called mixed boundary value problem different conditions are prescribed on different parts of the boundary Γ_0.

What condition should be prescribed on boundary of the holes Γ_i? This is the problem we want to discuss. Its solution depends on the physical background of the problem. We shall deal with two particular cases: the thermal conduction and torsion of a bar problem, where unusual boundary conditions appear.

3   Thermal Conduction Problem

We shall start with a usual interpretation of the Eq. (1). The domain Ω represents the volume of a two dimensional body (a thin plate thermally insulated on both faces), $u(x, y)$ is the temperature, $f(x, y)$ is a power of possible heat sources and $a(x, y)$ is the positive thermal conductivity coefficient, which is constant in case of a homogeneous isotropic material. In case of the Dirichlet boundary condition (2) the temperature u is given on the outer boundary Γ_0. The Neumann condition (3) prescribes the heat flux, particularly $\frac{\partial u}{\partial n} = 0$ means thermally insulated boundary.

The boundary condition on the holes depends on character of the holes. The hole is usually supposed to be thermally insulating and thus on the hole the thermal flux is zero, which leads to the homogeneous Neumann boundary condition

$$\frac{\partial u}{\partial n} = 0 \qquad \text{on } \Gamma_i, \quad i = 1, \dots, k . \tag{4}$$

The boundary value problem (1) with (2) and (4) is correctly formulated. Indeed, its weak formulation (with $\overline{u} = 0$) reads:

Find $u \in V$ such that

$$\iint_{\Omega} a \left(\frac{\partial u}{\partial x} \frac{\partial v}{\partial x} + \frac{\partial u}{\partial y} \frac{\partial v}{\partial y} \right) dx\,dy = \iint_{\Omega} fv\,dx\,dy \qquad \forall v \in V \; , \tag{5}$$

where $V = \{u \in H^1(\Omega),$ such that $u = 0$ on $\Gamma_0\}$ is a subspace of the Sobolev space $H^1(\Omega)$ of functions with square integrable derivatives. For a bounded coefficient $a(x, y) \geq \alpha > 0$ the left hand side of (5) is an elliptic bounded bilinear form on $V \times V$ and for $f \in L^2(\Omega)$ following the well known Lax-Milgram Lemma the problem admits unique solution. The same result can be proved for the case with nonzero Dirichlet boundary conditions $\overline{u} \in H^1(\Omega)$.

These boundary conditions on the holes can be obtained also by a limit procedure. Let us extend the Eq. (1) from Ω to Ω_0, i.e. into the holes, putting the conductivity coefficient inside the holes Ω_i to be a parameter q. Since the coefficient is the conductivity and the holes are considered to be an insulating material we put q tending to zero. In the thermal conduction problem with $u \in H^1(\Omega)$ besides the temperature u also the flux $g = -a \cdot \frac{\partial u}{\partial n}$ should be continuous (in sense of traces), thus the condition on Γ_i reads

$$a \cdot \left(\frac{\partial u}{\partial n} \right)_{\text{int}} = q \cdot \left(\frac{\partial u}{\partial n} \right)_{\text{ext}} \qquad \text{on } \Gamma_i \tag{6}$$

and for $q \to 0$ we obtain the Neumann condition $\left(\frac{\partial u}{\partial n} \right)_{\text{int}} = 0$.

Let us remark that other situation will be if the holes are supposed to be perfectly thermally conducting material. Then q tends to ∞, the temperature gradient ∇u tends to zero in the holes and we can assume the temperature u is constant in each hole $\overline{\Omega_i}$.

4   Torsion of a Bar

Different situation appears in the problem arising from mechanics: torsion of a bar. In the monographs on continuum mechanics, e.g. [BSS, K, NH], one can find that torsion of an elastic bar with simply connected profile Ω, i.e. occupying the volume $\Omega \times (0, \ell)$, is modelled by the Poisson equation

$$- \Delta u \equiv - \frac{\partial^2 u}{\partial x^2} - \frac{\partial^2 u}{\partial y^2} = 2 \qquad \text{in } \Omega \; . \tag{7}$$

The unknown u is the Airy stress function yielding the only nonzero components of the stress tensor $\tau = \{\tau_{ij}\}$:

$$\tau_{xz} = \tau_{zx} = \alpha \mu \frac{\partial u}{\partial y} \; , \qquad \tau_{yz} = \tau_{zy} = -\alpha \mu \frac{\partial u}{\partial x} \; , \tag{8}$$

where α is the torsion angle in radians and μ the sheer modulus of the material. The equation is completed with zero boundary condition $u = 0$ on the boundary Γ. The introduced boundary value problem considers only the simply connected profile of the bar, i.e. the profile without holes. To deal with a profile with holes we extend the derivation of the mathematical model.

Using the notation introduced in Sect. 2 we consider a bar $\Omega \times (0, \ell)$ with the profile Ω. The bar is fixed at the base $z = 0$ and the base $z = \ell$ is rotated along z-axis by an angle $\ell\alpha$. Let $v = (v_x, v_y, v_z)$ be the displacement vector. According to [NH] we adopt the hypothesis that in the x, y plane the cross-section $\Omega \times \{z\}$ rotates along the axis z as a rigid body and in z direction the cross-section is deflected by a function $\alpha\, \varphi(x, y)$. Then we can write

$$v_x = -\alpha\, y\, z\ , \qquad v_y = \alpha\, x\, z\ , \qquad v_z = \alpha\, \varphi(x, y)\ . \tag{9}$$

The corresponding strain tensor $e = \{e_{ij}\} = \frac{1}{2}\left(\nabla v + (\nabla v)^T\right)$ has components

$$e_{xz} = e_{zx} = \frac{1}{2}\,\alpha\left(\frac{\partial\varphi}{\partial x} - y\right), \qquad e_{yz} = e_{zy} = \frac{1}{2}\,\alpha\left(\frac{\partial\varphi}{\partial y} + x\right), \tag{10}$$

the other components $e_{xx}, e_{yy}, e_{zz}, e_{xy}, e_{yx}$ are zero. Simple computation yields

$$\frac{\partial e_{yz}}{\partial x} - \frac{\partial e_{xz}}{\partial y} = \frac{\alpha}{2}\left(\frac{\partial^2\varphi}{\partial x \partial y} + 1 - \frac{\partial^2\varphi}{\partial x \partial y} + 1\right) = \alpha. \tag{11}$$

The Hooke's law of linear elasticity with the sheer modulus μ yields

$$\tau_{xz} = 2\,\mu\, e_{xz} = \alpha\,\mu\left(\frac{\partial\varphi}{\partial x} - y\right), \qquad \tau_{yz} = 2\,\mu\, e_{yz} = \alpha\,\mu\left(\frac{\partial\varphi}{\partial y} + x\right), \tag{12}$$

all the other components $\tau_{xx}, \tau_{yy}, \tau_{zz}, \tau_{xy}, \tau_{yx}$ are zero.

The equilibrium equations $\sum_j \partial_j \tau_{ij} = f_i$ with zero forces $f_i = 0$ reduce to

$$\frac{\partial\tau_{xz}}{\partial x} + \frac{\partial\tau_{yz}}{\partial y} = 0, \qquad \frac{\partial\tau_{xz}}{\partial z} = 0, \qquad \frac{\partial\tau_{yz}}{\partial z} = 0. \tag{13}$$

The second and the third equality in (13) imply τ_{xz} and τ_{yz} are independent of z. The first equality ensures that the vector field $V = (-\tau_{xz}, \tau_{yz})$ is irrotational: $(\partial_x, \partial_y) \times (-\tau_{xy}, \tau_{yz}) = \partial_x \tau_{xz} + \partial_y \tau_{yz} = 0$.

Let us recall that for a simply connected domain Ω the irrotational vector field $V = (V_x, V_y)$ is potential, i.e. there exists a function u such that $\nabla u = V$. Thus (13) imply existence of a function u such that τ_{xz} and τ_{yz} are given by

$$\tau_{xz} = \alpha\,\mu\,\frac{\partial u}{\partial y}, \qquad \tau_{yz} = -\alpha\,\mu\,\frac{\partial u}{\partial x}. \tag{14}$$

If we express the components e_{xz} and e_{yz} using (12) by means of potential u and insert it into the Eq. (11), multiplying by $2/\alpha$ we obtain the equation

$$-\Delta u \equiv -\left[\frac{\partial^2 u}{\partial x^2} + \frac{\partial^2 u}{\partial y^2}\right] = 2 \qquad \text{in } \Omega. \tag{15}$$

The equation has to be completed with boundary conditions. Since zero surface forces are considered, on the boundaries Γ_i the traction vector $T = \tau \cdot n$ must have zero components T_x, T_y, T_z. Due to $n_z = 0$ the components T_x, T_y are zero. Inserting τ_{xz}, τ_{yz} from (14) to equality $T_z = \tau_{xz}n_x + \tau_{yz}n_y + \tau_{zz}n_z = 0$ we obtain

$$T_z = \tau_{xz}n_x + \tau_{yz}n_y = \tau_{xz}t_y - \tau_{yz}t_x = \mu\alpha\left(\frac{\partial u}{\partial y}\,t_y + \frac{\partial u}{\partial x}\,t_x\right) = \mu\,\alpha\,\frac{\partial u}{\partial t} = 0. \quad (16)$$

Therefore u is constant along each component Γ_i of the boundary. In the case of simply connected profile Ω the boundary $\partial\Omega = \Gamma_0$ has only one component and we can choose $u = 0$ on Γ_0.

5   Potentiality Conditions

Since we consider the profile Ω with holes, i.e. multiply connected domain Ω, the equalities (13) are necessary but not sufficient conditions for existence of a potential $u(x, y)$. Let us show a counterexample. Let us consider the vector field

$$V \equiv (V_x, V_y) = \left(\frac{-y}{x^2 + y^2}, \frac{x}{x^2 + y^2}\right)$$

on the ring shape domain $G = \{[x, y] \mid r^2 < x^2 + y^2 < R^2\}, \quad (0 < r < 1 < R)$. Simple calculation verifies $\partial_x V_y - \partial_y V_x = 0$, i.e. the vector field V is irrotational in G. But V has no potential on G. If there were a potential $V(x, y)$ then each line integral $\int_C (V_x\, dx + V_y\, dy)$ over a closed curve C in G would equal to zero. Let C be the unit circle $x = \cos s$, $y = \sin s$, $s \in \langle 0, 2\pi\rangle$. Then $x^2 + y^2 = 1$, $dx = -\sin s\, ds$, $dy = \cos s\, ds$ and

$$\int_C (V_x\, dx + V_y\, dy) = \int_0^{2\pi} (\sin^2 s + \cos^2 s)ds = \int_0^{2\pi} 1\, ds = 2\pi,$$

which contradicts potentiality.

To ensure that the irrotational vector field $V = (V_x, V_y)$ is potential we need to verify that its line integral is path independent, i.e. the line integral of V over any closed curve in Ω equals to zero. Since for each simply connected domain irrotational vector field V is potential, it is sufficient to test only the curves C which encircle the holes. In our case the vector field V can be continuously extended to the closure $\overline{\Omega}$, thus the conditions $\int_{\Gamma_i}(V_x\, dx + V_y\, dy) = 0$ ensure existence of the potential u. Using $dx = t_x\, ds$, $dy = t_y\, ds$ and relations $t_x = -n_y$ and $t_y = n_x$, the condition can be rewritten to

$$\int_{\Gamma_i} (-V_x\, n_y + V_y\, n_x)\, ds = 0 , \quad i = 1, \ldots, k. \quad (17)$$

In our problem condition (17) for ensuring existence of a potential u for vector field $(V_x, V_y) = (-\tau_{yz}, \tau_{xz})$ is satisfied since $\tau_{xz}n_x + \tau_{yz}n_y = T_z = 0$ on each Γ_i. Thus the potential $u(x, y)$ exists even for the profiles with holes.

Boundaries Γ_i are separated, thus we can choose $c_0 = 0$. On the other boundaries Γ_i there are conditions $u = c_i$ with undetermined constants $c_1, \ldots, c_k$.

According to the theory of elliptic partial differential equations for any choice of the constants $c_1, \ldots, c_k$ we obtain a solution u which yields different stress tensors τ. Since the solution of the real torsion problem should have unique solution, some additional conditions must be added.

What condition should be added? Reformulating the problem for the Airy stress function u we lost connection to the deflection function φ. The boundary value problem for u should be completed by a condition that the corresponding stress components τ_{xy}, τ_{xz} admit the deflection function φ. From (12) and (14) we have

$$\frac{\partial \varphi}{\partial x}(x, y) = \frac{\partial u}{\partial y}(x, y) + y, \qquad \frac{\partial \varphi}{\partial y}(x, y) = -\frac{\partial u}{\partial x}(x, y) - x. \qquad (18)$$

It is the problem of finding a potential φ from its differential $d\varphi = V_x \, dx + V_y \, dy$, where in this case

$$V_x = \frac{\partial u}{\partial y}(x, y) + y \ , \qquad V_y = -\frac{\partial u}{\partial x}(x, y) - x. \qquad (19)$$

Simple calculation with (15) verifies that the vector field (V_x, V_y) is irrotational:

$$\mathrm{rot}\, V = \frac{\partial V_x}{\partial y} - \frac{\partial V_y}{\partial x} = \frac{\partial^2 u}{\partial y^2} + 1 + \frac{\partial^2 u}{\partial x^2} + 1 = \Delta u + 2 = 0.$$

In our case of multiply connected domain Ω we need to add condition that each line integral is path independent, which is ensured by conditions (17). Equalities (18) yields

$$\int_{\Gamma_i} (-V_x n_y + V_y n_x) \, ds \equiv -\int_{\Gamma_i} \left(\frac{\partial u}{\partial y} n_y + \frac{\partial u}{\partial x} n_x \right) ds - \int_{\Gamma_i} (y \, n_y + x \, n_x) \, ds.$$

Integrand of the first integral equals to $\frac{\partial u}{\partial n}$. The second integral will be transformed using the Gauss-Ostrogradski theorem. Taking into account that our normal vector n is oriented inward the hole Ω_i we obtain

$$\int_{\Gamma_i} (x \, n_x + y \, n_y) \, ds = -\iint_{\Omega_i} \left(\frac{\partial x}{\partial x} + \frac{\partial y}{\partial y} \right) dx \, dy = -\iint_{\Omega_i} 2 \, dx \, dy = -2 \, |\Omega_i|,$$

where $|\Omega_i|$ means the area of the hole Ω_i. Thus with our orientation of the normal vector, the potentiality condition (17) yields the additional conditions

$$\int_{\Gamma_i} \frac{\partial u}{\partial n} \, ds = 2|\Omega_i|, \qquad i = 1, \ldots, k. \qquad (20)$$

Thus for the multiply connected profile Ω we have obtained:

Problem (P). *Find* $u \in C^2(\Omega) \cap C^1(\overline{\Omega})$ *and* $c_1, \ldots, c_k \in \mathbb{R}$ *such that*

$$
\begin{aligned}
-\Delta u &= 2 & &in\ \Omega, \\
u &= 0 & &on\ \Gamma_0, \\
u &= c_i & &on\ \Gamma_i,\ i = 1, \ldots, k, \\
\int_{\Gamma_i} \frac{\partial u}{\partial n} \, ds &= 2\,|\Omega_i| & &,\ i = 1, \ldots, k.
\end{aligned}
\qquad (21)
$$

Let us remark that as in the end of Sect. 3 also the torsion problem "with holes" can be obtained by limit procedure of problems "without holes" by the coefficient a extended to holes by $q \to \infty$.

6   Variational Formulation of the Torsion Problem

Let us derive Problem (P) from the variational formulation. The approach looks for a minimizer of the functional Φ over the set $\mathcal{U}$ of admissible potentials u.

An admissible potential u is defined on the domain Ω with holes $\Omega_1, \ldots, \Omega_k$. Since u has to be constant on Γ_i the constant can be extended inside the hole Ω_i to a function denoted again by u. Thus we can look for the potential u defined on Ω_0 being constant on the holes Ω_i with undetermined constants c_i.

Therefore let us define the set $\mathcal{U}$ of admissible potentials u by

$$\mathcal{U} = \left\{ u \in H^1(\Omega_0) \text{s. t. } u|_{\Gamma_0} = 0 \text{ and } u|_{\overline{\Omega}_i} = c_i, \ c_i \in \mathbb{R}, \ i = 1, \ldots, k \right\}. \quad (22)$$

Solution of $-\Delta u = 2$ minimizes the functional $\Phi(u) = \iint_\Omega [\frac{1}{2}(\nabla u)^2 - 2u] \mathrm{d}x \, \mathrm{d}y$.

Since the gradient of a constant function on Ω_i is zero on Ω_i, we extend the integral from Ω to the whole Ω_0. Then the variational formulation reads:

Problem (V). *Find $u \in \mathcal{U}$ such that it minimizes the functional*

$$\Phi(u) = \iint_{\Omega_0} \left[\frac{1}{2} \left(\frac{\partial u}{\partial x} \right)^2 + \frac{1}{2} \left(\frac{\partial u}{\partial y} \right)^2 - 2u \right] \mathrm{d}x \, \mathrm{d}y \quad (23)$$

over the set $\mathcal{U}$, i. e. the inequality $\Phi(u) \leq \Phi(v)$ holds for each $v \in \mathcal{U}$.

What is the relation between Problem (V) and Problem (P)?

Theorem. *Each sufficiently smooth solution of Problem (V) solves Problem (P).*

Sketch of the proof: Let the functional Φ attain its minimum at $u \in \mathcal{U}$. Since u is an inner element of $\mathcal{U}$ then for any $v \in \mathcal{U}$ the function ψ defined by $\psi(t) = \Phi(u + tv)$ attains its minimum at $t = 0$, and thus $\psi'(0) = 0$.

Computing $\psi'(t) = \frac{\mathrm{d}}{\mathrm{d}t} \Phi(u + tv)$ the condition $\psi'(0) = 0$ yields

$$\iint_{\Omega_0} \left[\frac{\partial u}{\partial x} \frac{\partial v}{\partial x} + \frac{\partial u}{\partial y} \frac{\partial v}{\partial y} - 2v \right] \mathrm{d}x \, \mathrm{d}y = 0 \qquad \forall v \in \mathcal{U}. \quad (24)$$

To obtain the result we shall use of the following well-known lemma:

Test lemma. *Let a continuous function f on a domain Ω satisfy*

$$\iint_\Omega f(x, y) \, v(x, y) \, \mathrm{d}x \, \mathrm{d}y = 0 \quad (25)$$

for each "test function" v from a set of functions on Ω containing for each open ball B in Ω a continuous function positive in the ball B and zero outside the ball B. Then the function f equals to zero in Ω.

To use the Test lemma we have to transform the equality (24) into the form (25). Assuming that u is twice differentiable, integration by parts, using $v = 0$ on Γ_0, $\Omega_0 = \Omega \cup \overline{\Omega_1} \cup \cdots \cup \overline{\Omega_k}$ and $\partial \Omega = \Gamma_0 \cup \Gamma_1 \cup \cdots \cup \Gamma_k$ the equality $\psi'(0) = 0$ can be rewritten in the form

$$\iint_\Omega \left(-\frac{\partial^2 u}{\partial x^2} - \frac{\partial^2 u}{\partial y^2} - 2 \right) v \, dx \, dy + \sum_{i=1}^{k} \left[\int_{\Gamma_i} \frac{\partial u}{\partial n} v \, ds - 2 \iint_{\Omega_i} v \, dx \, dy \right] = 0. \quad (26)$$

Choosing v which is nonzero only inside any ball $B \subset \Omega$, the Test lemma yields $-\Delta u - 2 = 0$ inside Ω, which yields the first equation of Problem (P).

Using the last obtained equality, the first integral in (26) vanishes. Let us take any function v such that $v = 1$ in a point in Ω_i and zero in the other holes. Then v equals to 1 on $\overline{\Omega_i}$. In this way we obtain

$$\int_{\Gamma_i} \frac{\partial u}{\partial n} v \, ds - 2 \iint_{\Omega_i} v \, dx \, dy = \int_{\Gamma_i} \frac{\partial u}{\partial n} \, ds - 2 \iint_{\Omega_i} 1 \, dx \, dy = \int_{\Gamma_i} \frac{\partial u}{\partial n} \, ds - 2|\Omega_i| = 0,$$

which implies the last condition of Problem (P). The remaining two conditions follow directly from the condition $u \in \mathscr{U}$ and its definition. Thus the differentiable solution u of Problem (V) is a solution of Problem (P).   □

7 Examples of Solutions to the Torsion Problem

Importance of the boundary conditions will be illustrated on an example for the torsion Problem (P) derived in Sect. 5. Let us consider a ring profile Ω with the outer radius R and the inner radius $r = \lambda R$ given by $r^2 < x^2 + y^2 < R^2$. In this case, see [FR], the solution u – the Airy stress function – is the same as in the case of the full circle $x^2 + y^2 < R^2$

$$u(x, y) = \frac{1}{2} \left(R^2 - x^2 - y^2 \right) \quad (27)$$

and due to rotational symmetry the corresponding deflection function φ is zero. Let us compare the solution u of this Problem (P) to the solution u^* of Problem (P*) on a "broken ring" $\Omega^* = \Omega \setminus (-R, -r) \times \{0\}$, see Fig. 2.

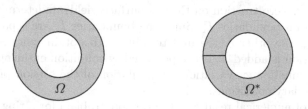

Fig. 2. Ring profile Ω and the "broken ring" profile Ω^*, $\lambda = 0.5$.

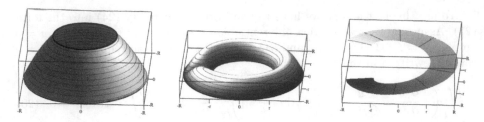

Fig. 3. Stress function u, u^* and the deflection φ^* (deflection φ is zero).

Boundary of Ω^* is already connected, thus Problem (P*) has zero condition $u^* = 0$ on the whole boundary $\partial\Omega^*$. The exact analytic solution can be computed by series method in polar coordinates, see [FR]. Graphs of the Airy stress functions u, u^* and the deflection function φ^* are on Fig. 3.

Important quantity is the profile moment J. It yields resistance of the twisted bar to the torque T by relation $T = \alpha\,\mu\,J$, where μ is the shear modulus and α the twisting rate. In the case of profile without holes it is $J = 2\iint_\Omega u(x,y)\,\mathrm{d}x\,\mathrm{d}y$. In [FR] the formula for profiles with holes is derived

$$J = 2\iint_\Omega u(x,y)\,\mathrm{d}x\,\mathrm{d}y + 2\sum_{i=1}^{k} c_i|\Omega_i|. \tag{28}$$

For the ring profile Ω the profile moment is $J = \frac{\pi}{2}R^4(1 - \lambda^4)$. Let us add some values computed in [FR]: If $\lambda = 0.5$ then the moment for ring profile is $J = 1.4726\,R^4$ while for the "broken ring" profile $J^* = 0.1844\,R^4$, for thinner ring $\lambda = 0.8$ the difference is even grater $J = 0.9274\,R^4$ and $J^* = 0.01476\,R^4$.

8   Conclusion

Modeling of bodies with holes leads to boundary value problems with boundary conditions on the holes. In case of thermal conduction equation with insulating holes the homogeneous Neumann condition can be prescribed. The result can be reached also by the limit procedure.

Modeling torsion of a bar with profile Ω leads to unusual boundary conditions on the holes. The Airy stress function u satisfies the Poisson equation $-\Delta u = 2$ in the profile Ω. Zero traction on the bar surface yields undetermined constant values c_i for u on boundaries Γ_i. Since the boundaries Γ_i are separated, $c_0 = 0$ can be chosen, the other c_i are undetermined. To obtain unique solution the integral condition is added. It is the potentiality condition ensuring existence of the deflection function φ. Variational formulation of the torsion problem yields the same boundary conditions.

Examples of numerical results to the torsion problem for a ring and "broken ring" profile from [FR] illustrates importance of the boundary conditions. Further examples of exact solution of the Torsion problem for various profiles with and without holes visualized by the MAPLE system can be found in [FNJ, FR].

Acknowledgments. This research is supported by Brno University of Technology, Specific Research project no. FSI-S-14-2290.

References

[BSS] Brdička, M., Samek, L., Sopko, B.: Mechanika Kontinua (Continuum Mechanics), in Czech, 3rd edn. Academia, Praha (2005)

[FNJ] Franců, J., Nováčková, P., Janíček, P.: Torsion of a non-circular bar. Eng. Mech. **19**, 45–60 (2012). http://www.engineeringmechanics.cz/pdf/19_1_045.pdf

[FR] Franců, J., Rozehnalová-Nováčková, P.: Torsion of a bar with holes. Eng. Mech. **22**, 3–23 (2015). http://www.engineeringmechanics.cz/pdf/22_1_003.pdf

[K] Kovář, A.: Theorie kroucení (Torsion theory), in Czech. Publ. House ČSAV, Praha (1954)

[L] Lanchon, H.: Torsion élastoplastique d'un arbre cylindrique de section simplement ou multiplement connexe. J. de Méchanique **13**, 267–320 (1974)

[NH] Nečas, J., Hlaváček, I.: Mathematical theory of elastic and elasto-plastic bodies: an introduction. In: Studies in Applied Mechanics, vol. 3. Elsevier Scientific Publishing Co., Amsterdam (1980)

[S] Sternberg, E.: On Saint-Venant torsion and the plane problem of elastostatics for multiply connected domains. In: Dafermos, C.M., Joseph, D.D., Leslie, F.M. (eds.) The Breadth and Depth of Continuum Mechanics, pp. 327–342. Springer, Berlin (1986)

Analysis of Model Error
for a Continuum-Fracture Model of Porous Media Flow

Jan Březina(✉) and Jan Stebel

Technical University of Liberec, Studentská 1402/2, 46117 Liberec, Czech Republic
{jan.brezina,jan.stebel}@tul.cz

Abstract. The Darcy flow problem in fractured porous media is considered. The fractures are treated as lower dimensional objects coupled with the surrounding continuum. Error estimates for the weak solution to such continuum-fracture model in comparison to the weak solution of the full model are derived. Validity of the estimates is inspected on one simple and one quasi-realistic case numerically.

Keywords: Darcy flow · Fractured media · Reduced model · Error estimate

1 Introduction

Deep subsurface deposits in a plutonic rock represent one of possible solutions for the final storage of radioactive waste. The primary reason is small hydraulic permeability of the bulk rock and thus slow migration of a possible leakage due to the ground water flow. On the other hand, granitoid formations contain fractures that may form a network of preferential paths with low volumetric water flow rate but with high velocity. The preferential paths pose a risk of fast transport of small amount of contaminant but in potentially dangerous concentrations. The large scale effect of the small scale fractures is challenging for numerical simulations since direct discretization requires highly refined computational mesh. One possible solution is to model fractures as lower dimensional objects and introduce their coupling with the surrounding continuum. A model for the saturated flow in the system matrix-fracture was formally derived in [7] by integrating the equations across the fracture. It was justified by an error estimate $O(\max\{h, \delta\})$, h being the mesh size and δ the fracture width, which holds inside the fracture for the solution of a particular mixed finite element approximation. The approach was then generalized by others e.g. to the case of curved fractures with variable width [1], non-matching grids [3] or to other equations or systems [4–6]. While most papers aim at the analysis or numerical solution of the continuum-fracture model, the precise statement declaring the relation of the original and reduced problem on the continuous level is, to our knowledge, missing. The presented estimates hold for the pressure gradient, which in turn controls the error in the velocity field which is required for the practical solute transport problems.

© Springer International Publishing Switzerland 2016
T. Kozubek et al. (Eds.): HPCSE 2015, LNCS 9611, pp. 152–160, 2016.
DOI: 10.1007/978-3-319-40361-8_11

In this paper we shall study the Darcy flow model, namely

$$\left.\begin{array}{rl}
\operatorname{div} \boldsymbol{q} = f & \text{in } \Omega, \\
\boldsymbol{q} = -\mathbb{K}\nabla p & \text{in } \Omega, \\
p = p_0 & \text{on } \partial\Omega,
\end{array}\right\} \tag{1}$$

where $\boldsymbol{q}$ is the Darcy flux, f is the source density, $\mathbb{K}$ is the hydraulic conductivity tensor, p is the piezometric head and p_0 is the piezometric head on the boundary. In what follows, $\Omega \subset \mathbf{R}^d$, $d = 2, 3$ will be a bounded domain with Lipschitz boundary (see Fig. 1, left), divided into the fracture

$$\Omega_f := \Omega \cap \left((-\delta/2, \delta/2) \times \mathbf{R}^{d-1} \right)$$

with thickness $\delta > 0$, and the surrounding set $\Omega_m := \Omega \setminus \overline{\Omega}_f$, called the matrix. The fracture interacts with the matrix on the interfaces

$$\gamma_1 := \Omega \cap \left(\{-\delta/2\} \times \mathbf{R}^{d-1} \right) \text{ and } \gamma_2 := \Omega \cap \left(\{\delta/2\} \times \mathbf{R}^{d-1} \right).$$

Normal vectors on these interfaces are denoted $\boldsymbol{n}_i$, $i = 1, 2$ with the orientation out of Ω_m. Further, we introduce the reduced geometry (see Fig. 1, right) where the fracture is represented by the manifold $\gamma := \Omega \cap \left(\{0\} \times \mathbf{R}^{d-1} \right)$ in its center. For a point $\boldsymbol{x} \in \mathbf{R}^d$, we shall write $\boldsymbol{x} = (x, \boldsymbol{y})^\top$, $\boldsymbol{y} \in \mathbf{R}^{d-1}$. For functions defined in Ω_f we define the tangent gradient along the fracture

$$\nabla_{\boldsymbol{y}} v := (0, \partial_{y_1} v, \ldots, \partial_{y_{d-1}} v)^\top,$$

and the average of v across the fracture:

$$\bar{v} := \frac{1}{\delta} \int_{-\delta/2}^{\delta/2} v(x, \cdot) \, dx.$$

We shall study the relation of (1) to the so-called *continuum-fracture model* on the reduced geometry:

$$\left.\begin{array}{rl}
-\operatorname{div}(\mathbb{K}\nabla p_m) = f & \text{in } \Omega_m, \\
p_m = p_0 & \text{on } \partial\Omega \cap \partial\Omega_m, \\
-\mathbb{K}\nabla p_m \cdot \boldsymbol{n}_i = q_i(p_m, p_f) & \text{on } \gamma_i, \ i = 1, 2, \\
-\operatorname{div}(\delta\mathbb{K}\nabla_{\boldsymbol{y}} p_f) = \delta\bar{f} + \displaystyle\sum_{i=1}^{2} q_i(p_m, p_f) & \text{in } \gamma, \\
p_f = p_0 & \text{on } \overline{\gamma} \cap \partial\Omega.
\end{array}\right\} \tag{2}$$

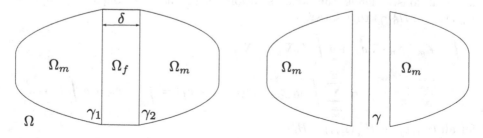

Fig. 1. The domain of the full model (left) and the reduced geometry (right).

The fluxes q_1, q_2 between the fracture and the matrix are given as follows:

$$q_i(v, w) := \frac{2\mathbb{K}_{|\gamma} \boldsymbol{n}_i \cdot \boldsymbol{n}_i}{\delta}(v_{|\gamma_i} - w_{|\gamma}), \; i = 1, 2,$$

where v and w are defined on $\overline{\Omega}_m$ and $\overline{\gamma}$ respectively. Our goal is to justify (2) as an approximation of (1) in the case of small δ. In particular, we shall prove that

$$\bar{u} - u_f \approx \delta \quad \text{and} \quad u_{|\Omega_m} - u_m \approx \delta^{3/2}$$

in a suitable sense.

The organization of the paper is as follows. In the next section we formulate and prove the main theoretical result on the error analysis. Then, in Sect. 3 we show numerical results which confirm the error estimates.

2   Asymptotic Properties of Continuum-Fracture Model

In what follows we assume that $\mathbb{K}$ is uniformly positive definite, bounded in Ω and has the following form:

$$\mathbb{K} = \begin{cases} \mathbb{K}_m & \text{in } \Omega_m, \\ \mathbb{K}_f = \begin{pmatrix} k_x & 0 \\ 0 & \mathbb{K}_{\boldsymbol{y}} \end{pmatrix} & \text{in } \Omega_f, \end{cases}$$

where $\mathbb{K}_f(x, \boldsymbol{y}) = \mathbb{K}_f(\boldsymbol{y})$. Further we consider right hand side $f \in L^2(\Omega)$. As the problem is linear, we can set $p_0 \equiv 0$ without loss of generality.

2.1   Weak Formulation

The ongoing analysis will be done in the framework of weak solutions. By $L^q(B)$ we denote the Lebesgue space on a measurable set B endowed with the norm $\| \cdot \|_{q,B}$, $q \in [1, \infty]$, $H^1(B)$ is the Sobolev space and $H_0^1(B)$ its subspace of functions with vanishing trace. We say that $p \in H_0^1(\Omega)$ is the weak solution of (1) if for every $v \in H_0^1(\Omega)$:

$$\int_\Omega \mathbb{K}\nabla p \cdot \nabla v = \int_\Omega fv. \tag{3}$$

Introducing the space

$$H_{bc}^1(\Omega_m) := \{v \in H^1(\Omega_m); \; v_{|\partial\Omega_m \cap \partial\Omega} = 0\},$$

we analogously define the weak solution of (2) as the couple $(p_m, p_f) \in H_{bc}^1(\Omega_m) \times H_0^1(\gamma)$ that satisfies

$$\int_{\Omega_m} \mathbb{K}_m \nabla p_m \cdot \nabla v_m + \delta \int_\gamma \mathbb{K}_{\boldsymbol{y}} \nabla_{\boldsymbol{y}} p_f \cdot \nabla_{\boldsymbol{y}} v_f$$

$$+ \sum_{i=1}^2 \int_{\gamma_i} q_i(p_m, p_f)(v_{m|\gamma_i} - v_f) = \int_{\Omega_m} fv_m + \delta \int_\gamma \bar{f} v_f \tag{4}$$

for all $(v_m, v_f) \in H_{bc}^1(\Omega_m) \times H_0^1(\gamma)$.

Let us remark that under the above assumptions on $\mathbb{K}$ and f, problems (3) and (4) have unique solutions.

2.2 Error Analysis of Asymptotic Model

Let $\sigma(\mathbb{A})$ denote the spectrum of a matrix $\mathbb{A}$. We use the following notation:

$$\underline{K}_m := \inf_{x \in \Omega_m} \sigma(\mathbb{K}_m(x)), \quad \underline{K}_y := \inf_{x \in \Omega_f} \sigma(\mathbb{K}_y(x)),$$

$$\underline{k}_x := \inf_{y \in \gamma} k_x(y), \quad \overline{k}_x := \sup_{y \in \gamma} k_x(y).$$

The main result of this section is the following error estimate.

Theorem 1. *Let $\delta > 0$, and assume in addition that the unique solution to (3) satisfies*

$$\partial_x^2 p \in L^q(\Omega_f) \text{ for some } q \in [2, \infty].$$

Then there is a constant $C := C(\Omega, \gamma) > 0$ independent of δ, $\mathbb{K}$ and f such that

$$\|\nabla_y(\bar{p} - p_f)\|_{2,\gamma} \leq C \sqrt{\frac{\overline{k}_x}{\underline{K}_y}} \|\partial_x^2 p\|_{q,\Omega_f} \delta^{1 - \frac{1}{q}}, \tag{5a}$$

$$\|\nabla(p - p_m)\|_{2,\Omega_m} \leq C \sqrt{\frac{\overline{k}_x}{\underline{K}_m}} \|\partial_x^2 p\|_{q,\Omega_f} \delta^{\frac{3}{2} - \frac{1}{q}}, \tag{5b}$$

$$\sum_{i=1}^{2} \|\bar{p} - p_{|\gamma_i} + p_{m|\gamma_i} - p_f\|_{2,\gamma} < C \sqrt{\frac{\overline{k}_x}{\underline{k}_x}} \|\partial_x^2 p\|_{q,\Omega_f} \delta^{2 - \frac{1}{q}}, \tag{5c}$$

where (p_m, p_f) is the solution of (4).

Proof. For any $\varepsilon \in (0, \delta/2)$ we define the sets

$$\Omega_{f\varepsilon} := \{(x, y) \in \Omega; \; -\delta/2 + \varepsilon < x < \delta/2 - \varepsilon\},$$
$$\Omega_{f\varepsilon}^- := \{(x, y) \in \Omega_f; \; x < -\delta/2 + \varepsilon\},$$
$$\Omega_{f\varepsilon}^+ := \{(x, y) \in \Omega_f; \; x > \delta/2 - \varepsilon\}$$

and an auxiliary operator $\Pi_\varepsilon : L^2(\Omega_m) \times L^2(\gamma) \to L^2(\Omega)$:

$$\Pi_\varepsilon(v_m, v_\gamma)(x, y) := \begin{cases} v_m(x, y) & \text{in } \Omega_m, \\ v_\gamma(0, y) & \text{in } \Omega_{f\varepsilon}, \\ \frac{1}{\varepsilon}(x + \frac{\delta}{2}) v_\gamma(0, y) - \frac{1}{\varepsilon}(x + \frac{\delta}{2} - \varepsilon) v_m(-\frac{\delta}{2}, y) & \text{in } \Omega_{f\varepsilon}^-, \\ -\frac{1}{\varepsilon}(x - \frac{\delta}{2}) v_\gamma(0, y) + \frac{1}{\varepsilon}(x - \frac{\delta}{2} + \varepsilon) v_m(\frac{\delta}{2}, y) & \text{in } \Omega_{f\varepsilon}^+. \end{cases}$$

Note that Π_ε maps $H_{bc}^1(\Omega_m) \times H_0^1(\gamma)$ into $H_0^1(\Omega)$. We use $v_\varepsilon := \Pi_\varepsilon(v_m, v_f)$, $v_m \in H_{bc}^1(\Omega_m)$, $v_f \in H_0^1(\gamma)$ as a test function in (3):

$$\int_{\Omega_m} \mathbb{K}_m \nabla p \cdot \nabla v_m + \int_{\Omega_{f\varepsilon}} \mathbb{K}_{\boldsymbol{y}} \nabla_{\boldsymbol{y}} p \cdot \nabla_{\boldsymbol{y}} v_f + \int_{\Omega_f \backslash \Omega_{f\varepsilon}} k_x \partial_x p \partial_x v_\varepsilon$$

$$+ \int_{\Omega_f \backslash \Omega_{f\varepsilon}} \mathbb{K}_{\boldsymbol{y}} \nabla_{\boldsymbol{y}} p \cdot \nabla_{\boldsymbol{y}} v_\varepsilon = \int_{\Omega_m} f v_m + \int_{\Omega_{f\varepsilon}} f v_f + \int_{\Omega_f \backslash \Omega_{f\varepsilon}} f v_\varepsilon. \quad (6)$$

Next we shall perform the limit $\varepsilon \to 0+$. Due to continuity of the integral we have:

$$\int_{\Omega_{f\varepsilon}} \mathbb{K}_{\boldsymbol{y}} \nabla_{\boldsymbol{y}} p \cdot \nabla_{\boldsymbol{y}} v_f \to \int_{\Omega_f} \mathbb{K}_{\boldsymbol{y}} \nabla_{\boldsymbol{y}} p \cdot \nabla_{\boldsymbol{y}} v_f = \delta \int_\gamma \mathbb{K}_{\boldsymbol{y}} \nabla_{\boldsymbol{y}} \bar{p} \cdot \nabla_{\boldsymbol{y}} v_f, \quad (7)$$

$$\int_{\Omega_f \backslash \Omega_{f\varepsilon}} \mathbb{K}_{\boldsymbol{y}} \nabla_{\boldsymbol{y}} p \cdot \nabla_{\boldsymbol{y}} v_\varepsilon \to 0, \quad (8)$$

$$\int_{\Omega_{f\varepsilon}} f v_f \to \int_{\Omega_f} f v_f = \delta \int_\gamma \bar{f} v_f, \quad (9)$$

$$\int_{\Omega_f \backslash \Omega_{f\varepsilon}} f v_\varepsilon \to 0, \ \varepsilon \to 0+. \quad (10)$$

The remaining term can be rewritten as follows:

$$\int_{\Omega_f \backslash \Omega_{f\varepsilon}} k_x \partial_x p \partial_x v_\varepsilon = \frac{1}{\varepsilon} \int_{\Omega_{f\varepsilon}^-} k_x \partial_x p (v_f - v_{m|\gamma_1}) - \frac{1}{\varepsilon} \int_{\Omega_{f\varepsilon}^+} k_x \partial_x p (v_f - v_{m|\gamma_2})$$

$$\to \sum_{i=1}^2 (-1)^{1+i} \int_\gamma k_x \partial_x p_{|\gamma_i} (v_f - v_{m|\gamma_i}), \ \varepsilon \to 0+. \quad (11)$$

Let $\boldsymbol{y} \in \mathbf{R}^{d-1}$ be fixed and define

$$P(x) := \frac{1}{\delta} \int_{-\delta/2}^x p(t, \boldsymbol{y}) \, dt.$$

Using the Taylor expansion

$$P(x) = P(-\delta/2) + (x + \frac{\delta}{2}) P'(-\delta/2) + \frac{(x + \frac{\delta}{2})^2}{2} P''(-\delta/2)$$

$$+ \frac{(x + \frac{\delta}{2})^2}{2} \int_{-\delta/2}^{\xi(x,\boldsymbol{y})} P'''(t) \, dt, \ \xi(x, \boldsymbol{y}) \in (-\delta/2, x), \quad (12)$$

we can show that

$$\bar{p}(\boldsymbol{y}) = P(\delta/2) = p(-\delta/2, \boldsymbol{y}) + \frac{\delta}{2} \partial_x p(-\delta/2, \boldsymbol{y}) + \frac{\delta}{2} \int_{-\delta/2}^{\xi(\delta/2,\boldsymbol{y})} \partial_x^2 p(t, \boldsymbol{y}) \, dt.$$

By a similar argument we obtain:

$$\bar{p}(\boldsymbol{y}) = p(\delta/2, \boldsymbol{y}) - \frac{\delta}{2} \partial_x p(-\delta/2, \boldsymbol{y}) + \frac{\delta}{2} \int_{\eta(-\delta/2,\boldsymbol{y})}^{\delta/2} \partial_x^2 p(t, \boldsymbol{y}) \, dt, \ \eta(x, \boldsymbol{y}) \in (x, \delta/2).$$

From this we can deduce that

$$\partial_x p_{|\gamma_i} = (-1)^{1+i} \left(\frac{2}{\delta}(\bar{p} - p_{|\gamma_i}) - \delta g_i \right), \quad i = 1, 2,$$ (13)

where

$$|g_i(\boldsymbol{y})| \leq \frac{1}{\delta} \int_{-\delta/2}^{\delta/2} |\partial_x^2 p(\cdot, \boldsymbol{y})|.$$ (14)

Summing up, (6)–(13) yields:

$$\int_{\Omega_m} \mathbb{K}_m \nabla p \cdot \nabla v_m + \delta \int_\gamma \mathbb{K}_{\boldsymbol{y}} \nabla_{\boldsymbol{y}} \bar{p} \cdot \nabla_{\boldsymbol{y}} v_f + \sum_{i=1}^2 \int_{\gamma_i} q_i(p, \bar{p})(v_{m|\gamma_i} - v_f)$$

$$= \int_{\Omega_m} f v_m + \delta \int_\gamma \bar{f} v_f + \delta \sum_{i=1}^2 \int_\gamma k_x g_i (v_{m|\gamma_i} - v_f).$$ (15)

Now we use $v_m := p - p_m$, $v_f := \bar{p} - p_f$ as test functions in (15) and (4), and subtract the resulting identities. We obtain:

$$\int_{\Omega_m} \mathbb{K}_m \nabla(p - p_m) \cdot \nabla(p - p_m) + \delta \int_\gamma \mathbb{K}_{\boldsymbol{y}} \nabla_{\boldsymbol{y}}(\bar{p} - p_f) \cdot \nabla_{\boldsymbol{y}}(\bar{p} - p_f)$$

$$+ \sum_{i=1}^2 \int_\gamma \frac{2k_x}{\delta} |p_{|\gamma_i} - p_{m|\gamma_i} - \bar{p} + p_f|^2 = \delta \sum_{i=1}^2 \int_\gamma k_x g_i (p_{|\gamma_i} - p_{m|\gamma_i} - \bar{p} + p_f).$$ (16)

Using Hölder's and Young's inequality we can estimate the right hand side of (16):

$$\delta \sum_{i=1}^2 \int_\gamma k_x g_i (p_{|\gamma_i} - p_{m|\gamma_i} - \bar{p} + p_f)$$

$$\leq \frac{\delta^{\frac{3}{2}}}{\sqrt{2}} \sum_{i=1}^2 \int_\gamma \sqrt{k_x} |g_i| \sqrt{\frac{2k_x}{\delta}} |p_{|\gamma_i} - p_{m|\gamma_i} - \bar{p} + p_f|$$

$$\leq \frac{\delta^3}{4} \bar{k}_x \sum_{i=1}^2 \|g_i\|_{2,\gamma}^2 + \frac{1}{2} \sum_{i=1}^2 \int_\gamma \frac{2k_x}{\delta} |p_{|\gamma_i} - p_{m|\gamma_i} - \bar{p} + p_f|^2.$$ (17)

From (14) and Hölder's inequality it follows that

$$\|g_i\|_{2,\gamma}^2 \leq \delta^{-\frac{2}{q}} |\gamma|^{\frac{q-2}{q}} \|\partial_x^2 p\|_{q,\Omega_f}^2.$$ (18)

Finally, (16), (17), (18) and the uniform positive definiteness of $\mathbb{K}$ yields:

$$\underline{K}_m \|\nabla(p - p_m)\|_{2,\Omega_m}^2 + \delta \underline{K}_{\boldsymbol{y}} \|\nabla_{\boldsymbol{y}}(\bar{p} - p_f)\|_{2,\gamma}^2$$

$$+ \frac{1}{\delta} k_x \sum_{i=1}^2 \|\bar{p} - p_{|\gamma_i} + p_{m|\gamma_i} - p_f\|_{2,\gamma}^2 \leq \frac{\bar{k}_x}{2} |\gamma|^{\frac{q-2}{q}} \|\partial_x^2 p\|_{q,\Omega_f}^2 \delta^{3-\frac{2}{q}},$$ (19)

from which the estimates (5) follow.

3   Numerical Experiments

In this section we present computational results that demonstrate the relevance of the continuum-fracture model in the discrete setting, in particular we study the dependence of the error between the full and the reduced model on the fracture thickness δ. For the numerical computation we used the mixed-hybrid FEM implemented in the code Flow123d [2].

For each δ, the solution to the continuum-fracture model computed on a sequence of meshes with different step h was compared either against the analytical solution of the full model or against a reference solution of the full model on a sufficiently fine mesh. This approach allows to distinguish the discretization error and the error of the reduced model.

3.1   Test 1 - Analytical Solution and Virtual Fracture

As the first, we consider a problem with the fixed constant conductivity $\mathbb{K} = \mathbb{I}$ admitting the exact solution $p^\delta(x, y) = e^x \sin y$ in the domain $\Omega := (-1, 1) \times (0, 1)$ with a virtual fracture $\Omega_f^\delta := (-\delta/2, \delta/2) \times (0, 1)$. In Fig. 2, we display the L^2 norm of the error in the pressure and in the velocity separately for the matrix and for the fracture domain. Numerical solution to the reduced model using $h \in \{0.04, 0.02, 0.01, 0.005, 0.0025\}$ is compared to the analytical solution. Estimated orders of convergence are consistent with the estimate (5), namely the convergence in the matrix domain is faster then in the fracture domain. All numerically estimated orders of convergence are higher then predicted by Theorem 1 since the solution is perfectly regular. Let us also note that $\|\partial_x^2 p^\delta\|_{\infty, \Omega_f}$, which stands on the right hand side of (5), is close to 1 for all δ. Unfortunately, the discretization error of the velocity on the matrix domain is only of the first order with respect to h which makes the numerical estimate of the order of convergence with respect to δ less precise.

3.2   Test 2 - Highly Permeable Fracture

In the second test we investigated more realistic case, with $\mathbb{K}_m = 1$ and $\mathbb{K}_f = 100$. We prescribed harmonic Dirichlet boundary condition $p^\delta(x, y) = \cos(x) \cosh(y)$ on the boundary of the domain $\Omega := (0, 2) \times (-0.5, 0.5)$ with the fracture along the line $x = 1$. Actually, we exploited symmetry of the problem, computing only on the top half and prescribing the homogeneous Neumann condition on the bottom boundary. In this case, no analytical solution is available, so for each δ we solved the full model on a mesh locally refined around the fracture and especially around its endpoint [1, 0.5], where the solution exhibits singularities. The corresponding continuum-fracture model was solved on a sequence of meshes for $h \in \{0.2, 0.1, 0.05, 0.02, 0.01, 0.005, 0.0025\}$.

As in the previous case, Fig. 3 displays the L^2–norm of the error in the pressure and in the velocity separately for the matrix and for the fracture domain. However, unlike in the previous case, the error is not consistent with the theoretical estimate. The error in the matrix domain is only of the order 1 with

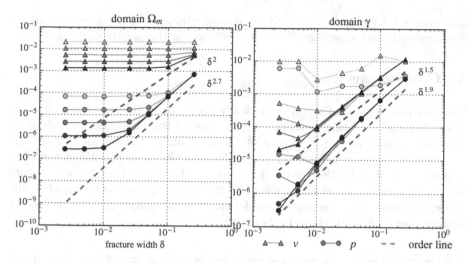

Fig. 2. Test 1 - virtual fracture $K_m = K_f = 1$, L^2 norm of the error in the pressure (p) and in the velocity (v), comparison against analytical solution. Each line connects values for the same h, darker colors correspond to smaller h.

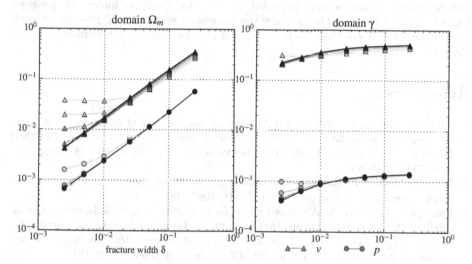

Fig. 3. Test 2 - permeable fracture, $K_m = 1$, $K_f = 100$, L^2 norm of the error in the pressure (p) and in the velocity (v), comparison against solution to the full model on refined mesh. Each line connects values for the same h, darker colors correspond to smaller h.

respect to δ and the error in the fracture domain stagnates. The main reason for this situation is the high and possibly unbounded value of the norm $\|\partial_x^2 p^\delta\|_{\infty,\Omega_f}$ as well as the norm $\|\partial_x^2 p^\delta\|_{2,\Omega_f}$, whose numerical estimates uniformly grow with decreasing h. On the other hand even if the norm of the second derivative is not bounded, we observe relatively good approximation of the pressure in both domains and good approximation of the velocity in the matrix domain.

4   Conclusions

We analyzed theoretically and numerically the error of the continuum-fracture model for the Darcy flow in a domain containing a fracture. The obtained error rates are related to the regularity of the solution to the full model and are in agreement with the result of [7]. Unfortunately, in applications the solution exhibits singularities at the intersection of the fracture boundary with the domain boundary which may lead to inaccuracy in the continuum-fracture model. Refinement or dedicated model may be necessary in order to obtain descent flux error at the fracture boundary.

Acknowledgement. This work was supported by the Ministry of Education, Youth and Sports under the project LO1201 in the framework of the targeted support of the "National Programme for Sustainability I" and the OPR & DI project Centre for Nanomaterials, Advanced Technologies and Innovation CZ.1.05/2.1.00/01.0005. Computational resources were provided by the MetaCentrum under the program LM2010005 and the CERIT-SC under the program Centre CERIT Scientific Cloud, part of the Operational Program Research and Development for Innovations, Reg. no. CZ.1.05/3.2.00/08.0144.

References

1. Angot, P., Boyer, F., Hubert, F.: Asymptotic and numerical modelling of flows in fractured porous media. ESAIM Math. Model. Numer. Anal. **43**(02), 239–275 (2009)
2. Březina, J., Stebel, J., Exner, P., Flanderka, D.: Flow123d 2011–2015. http:// flow123d.github.com
3. Frih, N., Martin, V., Roberts, J.E., Saâda, A.: Modeling fractures as interfaces with nonmatching grids. Comput. Geosci. **16**(4), 1043–1060 (2012)
4. Fumagalli, A., Scotti, A.: A reduced model for flow and transport in fractured porous media with non-matching grids. In: Cangiani, A., Davidchack, R.L., Georgoulis, E., Gorban, A.N., Levesley, J., Tretyakov, M.V. (eds.) Numerical Mathematics and Advanced Applications 2011, pp. 499–507. Springer, Heidelberg (2013)
5. Ganis, B., Girault, V., Mear, M., Singh, G., Wheeler, M.: Modeling fractures in a poro-elastic medium. Oil Gas Sci. Technol.-Rev. dIFP Energies nouvelles **69**(4), 515–528 (2014)
6. Lesinigo, M., D'Angelo, C., Quarteroni, A.: A multiscale Darcy-Brinkman model for fluid flow in fractured porous media. Numer. Math. **117**(4), 717–752 (2011)
7. Martin, V., Jaffré, J., Roberts, J.E.: Modeling fractures and barriers as interfaces for flow in porous media. SIAM J. Sci. Comput. **26**(5), 1667 (2005)

Probabilistic Time-Dependent Travel Time Computation Using Monte Carlo Simulation

Radek Tomis[✉], Lukáš Rapant, Jan Martinovič,
Kateřina Slaninová, and Ivo Vondrák

VŠB - Technical University of Ostrava, IT4Innovations,
17. listopadu 15/2172, 708 33 Ostrava, Czech Republic
{radek.tomis,lukas.rapant,jan.martinovic,
katerina.slaninova,ivo.vondrak}@vsb.cz

Abstract. This paper presents an experimental evaluation of probabilistic time-dependent travel time computation. Monte Carlo simulation is used for the computation of travel times and their probabilities. The simulation is utilizing traffic data regarding incidents on roads to compute the probability distribution of travel time on a selected path. Traffic data has the information about an optimal speed, a traffic incident speed and a probability of a traffic incident to occur. The exact algorithm is used for the comparison of the simulation.

Keywords: Probabilistic time-dependent route planning · Monte Carlo simulation · Probabilistic speed profiles · Uncertainty · Traffic incidents

1 Introduction

Finding the fastest path in the road network is a common task for navigation services. The efficient route planning can lower travel time and therefore lower travel costs for drivers. Some navigation services use traffic data on road networks to compute more accurate paths. However, most navigation services use only average data of historic travel times on roads for computation of the paths, even though this is not sufficient in many cases. Some roads can be for example congested at the time of rush hours, which leads to larger delays, or there can be traffic incidents on the roads.

Departure time highly affects route planning in the real road network. The fastest path is actually a time-dependent shortest path [12] in a graph with variable travel times used as edge weights. Travel times on the edges are usually computed as historic averages [14], even though this is not sometimes sufficient [5,8].

The main disadvantage of time-dependent route planning is that it does not take accidental events into account. There can be irregular but recurrent traffic congestions at some roads which navigation services should consider in the computation of the path and offer alternative routes [10]. Even though the probability of the traffic congestion can be very low, the delay in the case the

© Springer International Publishing Switzerland 2016
T. Kozubek et al. (Eds.): HPCSE 2015, LNCS 9611, pp. 161–170, 2016.
DOI: 10.1007/978-3-319-40361-8_12

congestion happens can be very long. Therefore, we have to consider uncertain traffic events and their probabilities in the computation of the path [11,17].

Traffic incidents and congestions cause uncertainty in travel times. There are existing algorithms [2,3,9,15], which try to find a solution for shortest path problem in uncertain networks. These algorithms are very time demanding even with very small road networks, because they use probability directly in the computation of the shortest path. The computation time is long even with the use of heuristics and reaches tens of seconds [9]. Example of such algorithm is P* [9]. P* uses static and time-independent road network, which is not desirable in our case, because we aim to take advantage of time-dependency of travel times in our algorithm. P* algorithm searches the solution for top-k path query, which results in paths with arrival probability or travel time better then some specified threshold value. In contrast to P*, we are interested in obtaining the complete probability distributions of travel time on paths. Therefore, we proposed a new approach to route planning with uncertainty [16] based on the computation of full probability distributions of travel time on some selected paths between two locations. Probabilistic travel times on these paths can be analyzed and the best path can be recommended based on the results. The selection of these paths can be done in many ways, for example by the computation of alternative routes [1,4].

In this paper, a Monte Carlo simulation is used for the computation of probabilistic travel times for a given path. The simulation randomly selects the state of the traffic on individual road segments using historical traffic data and then computes the travel time at the end of the path. The simulation has to be performed repeatedly to obtain the result in the form of the probability distribution of travel time. The accuracy of the result is dependent on the number of simulation iterations. The description of the simulation is presented at Sect. 2. Section 3 then presents the experiments and the description of traffic data.

2   Probabilistic Travel Time Computation

Road network is composed of junctions and road segments. Road segments connect junctions and have additional attributes associated with them, e.g. length, road category, capacity, etc. Road network can be represented as a directed graph $G = (V, E, w)$ defined as a set of vertices V, a set of edges E and function w, which assigns probabilistic speed profiles to edges. Vertices represent junctions, edges represent road segments and edge e_{ij} connects a pair of vertices v_i and v_j.

Probabilistic speed profile has information about time-dependent speed and its probability on related road segment, which means there is a different set of speeds with probabilities for every edge depending on current time. Function w returns a probabilistic speed profile as an ordered pair of speed s and probability p and is defined as $w(e, t, y) = (s, p)$, where $e \in E$, $t \in N$ is a number of time window and $y \in \{1, 2, \ldots Y\}$ is a number of specific profile type (e.g. 1 indicates free flow profile and other numbers are for profiles with different levels of traffic). The argument t denotes time window of profiles, so for interval length I, the time window is $\langle t \cdot I; (t+1) \cdot I \rangle$. The sum of the probability for one edge and one time window across all profile types has to be 1, or in other words $\sum_{y=1}^{Y} w(e,t,y) =$ $(a,1)$, where $a \in \mathbb{R}^+$ is the sum of speeds across the all profile types.

The storage for probabilistic speed profiles can be realized as a multidimensional sparse array, because all arguments of w are integers (edge can be represented as two integers of corresponding vertex numbers).

Probabilistic speed profiles can be created from various traffic data. There may be different profile types for variety of situations, e.g. free flow, congestions, traffic jams. These profiles can be created in many ways. One of them uses LoS (Level-of-Service - the measured speed divided by the normal free flow speed). Figure 1 shows probability intensity computed from values under 50 % LoS (traffic problems).

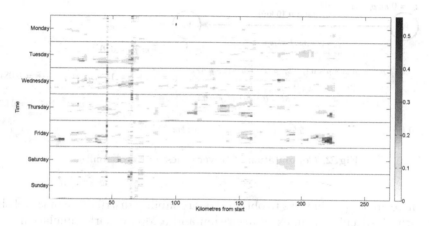

Fig. 1. Probability intensity for values < 50 % LoS

Our approach to probabilistic route planning [16] is based on the computation of the probability distribution of travel time on paths between source and destination locations. This paper is focused only on the computation of the probability distribution of travel time at the end of some selected path composed as the line of road segments (route) for specified departure time t_d. Let us denote this path $P_n = (v_0, v_1, \ldots, v_n)$, where $v_i \in V$, indices $i \in \{0, \ldots, n\}$ denote sequence of vertices in the path P_n, n denotes number of vertices in P_n, and $P_n \subseteq G$. The path starts at a vertex v_0 and ends at a vertex v_n. The probability distribution of travel time can be computed at any arbitrary vertex v_i of P_n with the knowledge of probabilistic speed profiles on corresponding edges. However, we are usually interested only in the probability distribution of travel time of the entire path P_n, which is the result at the last vertex v_n.

The example of the computation of travel times and probabilities at the end of one edge of P_n is presented in Fig. 2. Let us have probabilistic speed profiles $S_1^{t_1}$, $S_2^{t_1}$, $S_1^{t_2}$ and $S_2^{t_2}$ represented as pairs of speed and probability (s, p), where superscripts t_1 and t_2 represent the first and the second time window and subscripts 1 and 2 represent free flow and congested flow, respectively. The profile time window length is set to 15 min. The input at vertex v_0 is the departure time t_d and is set at 0 min. The output are pairs of current travel time and corresponding probability of that travel time (t, p_t). If the computation for profile $S_j^{t_1}$, $j \in \{1, 2\}$ exceeds the boundaries of time window in the middle of the edge, the profiles for the next time window (i.e. $S_j^{t_2}$, $j \in \{1, 2\}$) has to be used for the rest of the traversed edge (including the possibility of changing the type of profile j). Therefore in this case, there are three resulting pairs as the output at the end of the edge e_{01} representing all possible combinations of traffic levels during the traversing time. These results can be used for the computation of the probability distribution of travel time at vertex v_1.

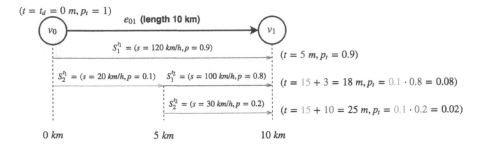

Fig. 2. Computation of travel times and probabilities

There are two main ways to obtain this probability distribution of travel time for path P, specifically an exact algorithm and a Monte Carlo simulation.

2.1   Exact Algorithm

The exact computation can be done using all possible combinations of probabilistic speed profiles on the path P_n. Travel times and their probabilities grow as a tree with the root node represented by v_0, leaf nodes represented by v_n, and each path between v_0 and v_n represents one specific combination of probabilistic speed profiles. The result of this algorithm is a complete probability distribution of travel time. Algorithm 1 shows the pseudo code for the exact tree computation. We use this algorithm as a baseline for evaluation of Monte Carlo simulations. This algorithm was already presented in greater depth in our previous article [16].

The tree can be very large; therefore there can be any number of nodes in between v_0 and v_n. The exact number of nodes depends on a quantity of probabilistic speed profiles and also on the departure time, because our tree can be additionally forked between two neighbour nodes if travel time at the end of the belonging edge would exceed the specific time window of the current probabilistic speed profile. Therefore, the time complexity is the main issue with this approach.

Algorithm 1. Compute Exact Probability Distribution of Travel Time for Given Array of Edges and Departure Time

---

1: **procedure** EXACTALGORITHM(*edges*, t_d)
2:     push to stack {*edge* = *edges*[0], *time* = t_d, *probability* = 1}
3:     **while** stack is not empty **do**
4:         *actual* = pop from stack
5:         *edgeResults* = GetTravelTimeProbabilities(*actual*)
6:         **for** *r* in *edgeResults* **do**   ▷ *r* is a pair of *time* and *probability*
7:             **if** *actual* → *edge* is last edge **then**
8:                 add *r* to result
9:             **else**
10:                push to stack {*actual* → *nextEdge*, *r* → *time*, *r* → *probability*}
11:            **end if**
12:        **end for**
13:    **end while**
14: **end procedure**

2.2   Monte Carlo Simulation

Monte Carlo simulation is another approach to obtain the result. Monte Carlo simulation is a computerized mathematical technique that allows to simulate and describe the behavior of complex system for purposes of risk analysis and decision making (more thorough description can be found here [13]). The technique is used in widely disparate fields as economy, engineering and environmental modeling. During a Monte Carlo simulation, input values are randomly sampled from the input probability distributions. Each set of these samples is then used for deterministic computation of the system. This process is called simulation, and its resulting outcome is recorded. Monte Carlo simulation repeats this simulation hundreds or thousands of times. It is logical that the quality of results improves with number of simulations performed. The resulting probability distribution is the probability distribution of possible outcomes of the system. Monte Carlo simulation therefore provides a comprehensive view of what may happen and how likely it is to happen. Progress of Monte Carlo simulation can be summarized into the following steps:

- Define a domain of possible inputs and number of simulations.
- Generate the inputs randomly from their probability distribution.
- Perform a deterministic computation on these inputs (simulations).
- Aggregate and evaluate the results.

Monte Carlo simulation has a number of advantages. Its results are easily interpretable and, providing that enough simulations are done, they represent the probabilistic behavior of the simulated system. This method is also basically very fast. Its only limits are how fast can we generate random inputs and how fast can we do the deterministic simulations. Generating random numbers may seem trivial, but Monte Carlo methods require large amounts of random numbers. For example, our simulations are not that large as each requires hundreds of inputs at worst. However, when we take into account that we perform these simulations several thousand times and that we want to run Monte Carlo simulations several times with slightly different inputs, we can easily get to hundreds of millions random numbers required for computations even on a relatively simple problem.

Our algorithm is simulating traversal of the tree from v_0 to v_n by randomly selecting probabilistic speed profiles on the edges of P_n, and computing travel time of the path. The simulation is repeated for a specified number of samples, and then the probability is created from computed travel times. See Algorithm 2 for pseudo code of our algorithm. The result does not have to be exact, because there is a random factor. However, the accuracy should be improved with greater number of samples.

Algorithm 2. Monte Carlo Simulation for Probabilistic Travel Time Computation for Given Array of Edges, Departure Time and Number of Samples

---

1: **procedure** MONTECARLOSIMULATION(*edges*, t_d, *samples*)
2:     **for all** *samples* **do**
3:         $t = t_d$
4:         **for** e in *edges* **do**
5:             r = generate random number between 0 and 1
6:             t = ComputeTravelTime(e, t, r)   ▷ speed profile is selected by r
7:         **end for**
8:         add travel time t to result
9:     **end for**
10: **end procedure**

Speed of deterministic simulation can be influenced by the use of parallelization, because individual simulations are strictly independent of each other and can be run in parallel. The scaling of the simulation parallelization should be almost ideal. On the other hand, the parallelization of the exact algorithm is much more complicated, because the algorithm needs to know the probability distribution of travel time at the end of the previous road segment before it can compute the probability distribution of travel time at the end of the following road segment. Therefore the exact algorithm involves dependent tasks and the parallelization should not scale very well.

3   Experiment

The experimental results of our algorithm are presented in this section. We use the road network created from Traffic Message Channel (TMC) segments in the Czech Republic, because it covers the backbone of the real road network.

We used probabilistic speed profiles computed from traffic data obtained due to online traffic monitoring system viaRODOS [6]. The experimental set of data was extracted for two months (October and November 2014). Probabilistic speed profiles were created for interval length $I = 15$ minutes for each day of the week and each measured place. Two profile types were created. The first profile (free flow) was defined as an average speed and its probability from values equal or higher than 50 percent of LoS, and the second profile (traffic problems) was defined as an average speed and its probability from the values under 50 percent of LoS.

The path from Praha to Brno along the highway D1 in the Czech Republic was chosen for the experiment. This collection was already used in our previous article [16], because it contains high number of incidents, and is more challenging to compute than collection with fewer incidents. The whole path is composed of 84 road segments (edges). The departure time was set at 2 pm on Wednesday, because there are many traffic incidents on the highway D1 at that time. Figure 3 shows the exact probability distribution for the specified settings. Table 1 shows the comparison between computation time of Monte Carlo simulation and the exact algorithm for different number of road segments. Only one processor core of Intel Sandy Bridge E5-2665 at 2.4 GHz was used for this experiment. As we can see, the computation time of the exact algorithm is increasing rapidly with greater number of road segments, while the time for Monte Carlo simulation is increasing in linear fashion.

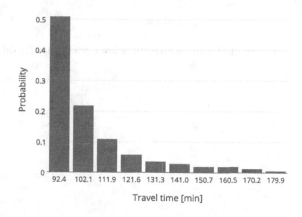

Fig. 3. Probability distribution of travel time

Table 1. Comparison of Computation Times

		Computation time [s]			
		Exact	Monte Carlo samples		
			10^3	10^4	10^5	10^6
Number of Segments	40	0.700	0.027	0.205	2.050	19.969
	50	7.700	0.031	0.244	2.445	24.651
	60	122.000	0.037	0.301	2.905	28.918
	70	667.000	0.038	0.352	3.510	34.139
	80	4849.000	0.040	0.383	3.837	38.250

The accuracy of the simulation compared to the exact algorithm is presented in Fig. 4. Each mean absolute error (MAE) of travel time was computed for 100 Monte Carlo simulations with specified number of samples. MAE was computed as follows:

$$\text{MAE} = \frac{1}{n} \sum_{i=1}^{n} |f_i - y|,$$

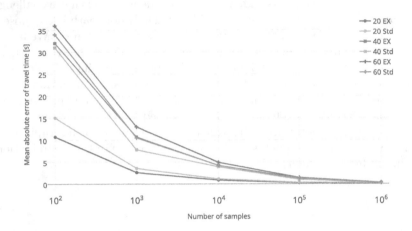

Fig. 4. Mean absolute error of Monte Carlo simulation travel times for different number of segments

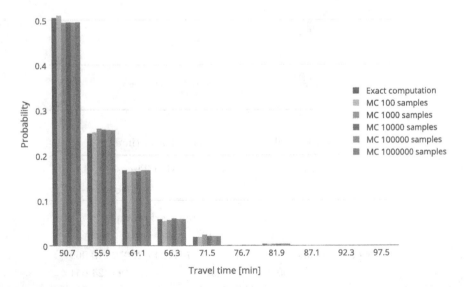

Fig. 5. Comparison of histograms of exact algorithm and Monte Carlo simulation for 40 road segments

where f_i are individual results of Monte Carlo simulations and y is true value obtained from the exact algorithm. There are always two different lines for each number of road segments (the first number in the legend). The first line is computed for expected values (EX) and the second for standard deviations (Std). As expected, the accuracy is higher with greater number of samples. The comparison between probability distributions of exact algorithm and Monte Carlo simulation with different number of samples is presented in Fig. 5. We can see that the histograms of the probability distribution of Monte Carlo simulations follow the histogram of the exact algorithm very precisely. In the case of million of simulations, the Earth Mover's distance of the distributions is 0.02, which means almost perfect match (Earth Mover's distance is described in [7]).

4 Conclusion

The Monte Carlo simulation for computation of the probability distribution of travel time on the selected path was presented in this paper. The simulation proved to be a suitable substitute for the exact algorithm. While the computational time of the exact algorithm is increasing very steeply, the computational time of Monte Carlo simulation is increasing in linear fashion. This allows to compute the results for a greater number of road segments, which would be almost impossible for the exact algorithm. The accuracy of the results of Monte Carlo simulation depends on the number of samples. It is relatively precise, even with lower numbers of samples.

Acknowledgment. This work was supported by The Ministry of Education, Youth and Sports from the National Programme of Sustainability (NPU II) project "IT4Innovations excellence in science - LQ1602", supported by the internal grant agency of VŠB Technical University of Ostrava, Czech Republic, under the project no. SP2015/157 "HPC Usage for Transport Optimisation based on Dynamic Routing" and data were provided by "Transport Systems Development Centre" co-financed by Technology Agency of the Czech Republic (reg. no. TE01020155).

References

1. Abraham, I., Delling, D., Goldberg, A.V., Werneck, R.F.: Alternative routes in road networks. J. Exp. Algorithmics **18**, 1.3:1.1–1.3:1.17 (2013)
2. Chen, B., Lam, W., Sumalee, A., Li, Q., Shao, H., Fang, Z.: Finding reliable shortest paths in road networks under uncertainty. Netw. Spat. Econ. **13**(2), 123–148 (2013)
3. Chen, B.Y., Lam, W.H.K., Sumalee, A., Li, Q., Tam, M.L.: Reliable shortest path problems in stochastic time-dependent networks. J. Intell. Transp. Syst. **18**(2), 177–189 (2014)
4. Dees, J., Geisberger, R., Sanders, P., Bader, R.: Defining and computing alternative routes in road networks. CoRR, abs/1002.4330 (2010)
5. Fan, Y., Kalaba, R., Moore I, J.E.: Arriving on time. J. Optim. Theory Appl. **127**(3), 497–513 (2005)

6. Fedorčák, D., Kocyan, T., Hájek, M., Szturcová, D., Martinovič, J.: viaRODOS: monitoring and visualisation of current traffic situation on highways. In: Saeed, K., Snášel, V. (eds.) CISIM 2014. LNCS, vol. 8838, pp. 290–300. Springer, Heidelberg (2014)
7. Ling, H., Okada, K.: An efficient earth mover's distance algorithm for robust histogram comparison. IEEE Trans. Pattern Anal. Mach. Intell. (2007)
8. Hofleitner, A., Herring, R., Abbeel, P., Bayen, A.: Learning the dynamics of arterial traffic from probe data using a dynamic bayesian network. IEEE Trans. Intell. Transp. Syst. 13(4), 1679–1693 (2012)
9. Hua, M., Pei, J. Probabilistic path queries in road networks: traffic uncertainty aware path selection. In: Proceedings of the 13th International Conference on Extending Database Technology, pp. 347–358. ACM (2010)
10. Miller-Hooks, E.: Adaptive least-expected time paths in stochastic, time-varying transportation and data networks. Networks 35–52 (2000)
11. Nikolova, E., Brand, M., Karger, D.R.: Optimal route planning under uncertainty. In: ICAPS, vol. 6, pp. 131–141 (2006)
12. Orda, A., Rom, R.: Shortest-path and minimum-delay algorithms in networks with time-dependent edge-length. J. ACM 37(3), 607–625 (1990)
13. Kroese, D.P., Rubinstein, R.Y.: Simulation and the Monte Carlo Method. Wiley-Interscience, Hoboken (2007)
14. Rice, J., Van Zwet, E.: A simple and effective method for predicting travel times on freeways. Intell. Transp. Syst. 5(3), 200–207 (2004)
15. Sun, S., Duan, Z., Sun, S., Yang, D.: How to find the optimal paths in stochastic time-dependent transportation networks? In: 2014 IEEE 17th International Conference on Intelligent Transportation Systems, pp. 2348–2353. IEEE (2014)
16. Tomis, R., Martinovič, J., Rapant, L., Slaninová, K., Vondrák, I.: Time-dependent route planning for the highways in the Czech Republic. In: Saeed, K., Homenda, W. (eds.) CISIM 2015. LNCS, vol. 9339, pp. 145–153. Springer, Berlin (2015)
17. Yang, B., Guo, C., Jensen, C.S., Kaul, M., Shang, S.: Multi-cost optimal route planning under time-varying uncertainty. In: Proceedings of the 30th International Conference on Data Engineering (ICDE), Chicago, IL, USA (2014)

Model of the Belousov-Zhabotinsky Reaction

Dalibor Štys[(✉)], Tomáš Náhlík, Anna Zhyrova, Renata Rychtáriková,
Štěpán Papáček, and Petr Císař

Faculty of Fisheries and Protection of Waters,
South Bohemian Research Center of Aquaculture and Biodiversity of Hydrocenoses,
Institute of Complex Systems, University of South Bohemia in České Budějovice,
Zámek 136, 373 33 Nové Hrady, Czech Republic
stys@jcu.cz
http://www.frov.jcu.cz/cs/ustav-komplexnich-systemu-uks

Abstract. The article describes results of the modified model of
the Belousov-Zhabotinsky reaction which resembles well the limit set
observed in an experiment in the Petri dish. We discuss the concept of the
ignition of circular waves and show that only an asymmetrical ignition
leads to the formation of spiral structures. From the qualitative assumptions on the behavior of dynamic systems we conclude that the reactants
in the Belousov-Zhabotinsky reaction likely forms a regular grid.

Keywords: Chemical self-organization · Multilevel cellular automata ·
Spiral formation

1 Introduction

The Belousov-Zhabotinsky (B-Z) reaction [1–3] is an experimentally easily accessible example of chemical self-organization. Many chemical reactions which con-
tribute to this process are already known, e.g. [4], and new chemical reactions
and compounds are being added every year (e.g., [5]).

Until recently, the reaction-diffusion simulations based on PDEs, which are
central models of chemical processes, have been used extensively to explain the
self-organizing Belousov-Zhabotinsky reaction [6]. This kind of simulation expects
an instantaneous chemical change. However, in reality, any chemical reaction is a
combination of the physical breaking of chemical bonds, splitting, and diffusion of
new chemical moieties over a time span – a defined elementary timestep needed for
the progression of the spatial-limited chemical process. Therefore, for the description of the B-Z reaction, we chose a kind of cellular automaton – a hodgepodge
machine, see e.g. [7,8] – for the time-spatially discrete simulation.

The hodgepodge machine is able to correctly simulate the final state of the
reaction course of the B-Z reaction as well as the formation of the mixture of spirals and waves. According to Wuensche [9], the final state is considered to be the
limit set and the course of the experiment is the trajectory through the state space.
In this way, all events which precede the establishment of spirals and waves are

© Springer International Publishing Switzerland 2016
T. Kozubek et al. (Eds.): HPCSE 2015, LNCS 9611, pp. 171–185, 2016.
DOI: 10.1007/978-3-319-40361-8_13

state space trajectories by which the system travels through ro reach the ergodi-
cally behaving dynamic process of the limit set.

Hodgepodge machine NetLogo algorithm – standard steps [10]

```
1   ;; all calculations are made in int
2
3   const
4       maxState; the maximal state level
5       k1 ;; the weight of the cell in the ignition state from the
              interval (0, maxState)
6       k2 ;; the weight of the cell with ignition state maxState
7       g ;; the number of levels crossed in one simulation step
8
9   patch-own
10      state_{x,y}[t] ;; the state level of the patch in step t
11      a ;; the number of the cells in the Moore neighborhood in
              the state from the interval (0, maxState)
12      b ;; the number of cells in the Moore neighborhood in
              maxState
13
14  to setup
15    foreach x,y
16    state_{x,y} = random(maxState + 1)
17  end
18
19  begin
20    foreach x,y
21    count a
22    count b
23    ifelse state_{x,y}[t] = maxState
24    state_{x,y}[t+1] = 0
25      ifelse state_{x,y}[t] = 0
26    state_{x,y}[t+1] = int(a/k1) + int(b/k2)
27      state_{x,y}[t+1] = int((sum [state] of neighbors)/(a+b+1))
              +g)
28        if state_{x,y}[t+1] > maxState
29              state_{x,y}[t+1] = maxState
30  end
```

In brief, we consider four processes in the hodgepodge machine algorithm
(see above) which describe the behavior of the chemical reaction at the quantum
(electron) scale:

1. Deexcitation – phase transition – bond breakage upon reaching the maximal
   level (lines 23–24 and 28–29),
2. spreading the "infection" from neighboring cells (lines 25–26),
3. excitation process which is simulated by the acquisition of cell levels (energy
   transfer) from the average of neighboring cells (line 27), and
4. growth of the excitation inside the cell (line 27).

Items 1–2 are the most speculative processes, but we have at least the experimental evidence for them. Further, we assume that processes 3 and 4 occur on

the border and inside of the structural element, respectively. In this aspect, the hodgepodge machine may be considered as the most elementary example with (a) only forward (excitation) processes and (b) a linear increase of the process inside the spatial element and on its border. The automaton depends upon 4 parameters, $maxState$, k_1, k_2, g, as well as on initial (an ignition phase) and boundary conditions (usually periodic).

In this article we report some experimental modifications on the hodgepodge machine. We started with only a few ignition points and varied the number of states and the constants g and $maxState$. Constants k_1 and k_2 are critical for the course of the system dynamics. If $\frac{1}{k_1} + \frac{0}{k_2} < 0.5$ and $\frac{0}{k_1} + \frac{1}{k_2} < 0.5$, we observed a limit set of waves and spirals. In all other cases, we observed filaments. We examine the influence of constants k_1, k_2, g, and $maxState$.

2   Methods

We used the Wilensky NetLogo modeling system [10] for its versatility and, in the discussion, Wuensche's terminology [9]. Our proposed modification of the Wilensky implementation is as follows:

Hodgepodge machine NetLogo algorithm – exponential start [11]

```
. . . . . . . . . .

const
      . . . . . .

      meanPosition
      ;; the mean of the random exponential distribution of the
            starting state (a new input constant)

. . . . . . . . . .

setup
   foreach x, y
   state_{x,y} = random−exponential ((maxState + 1) *
      meanPosition)                              ;; modified l. 16
   end
. . . . . . . . . . . .

begin
   foreach x,y

. . . . . . . . .

      state_{x,y}[t+1] = round(a/k1 + b/k2)      ;; modified l. 26

. . . . . . . . . . .

end
```

As shown, the Wilensky model was modified in two ways. The first change (line 16) was introduced to ensure that the simulation started with a small number of ignition points, similar to those found in the experiment. By a proper set-up of the constants $maxState$ and $meanPosition$, we achieved the condition in which the majority of points are in the initial $state_{x,y}$ higher than $maxState$. In the next step, we obtained a very few points in the non-zero state. This modification allowed us to observe the early phase of the wave formation in a way different from that of the original hodgepodge machine, but more similar to the evolution observed in the experiment.

The second modification (line 26) was implemented in order to allow us to start from these few ignition points. However, the non-intentional result of this modification was the increased similarity between the experiment and its simulation (model) (Fig. 1), which illustrates the importance of this ignition rule.

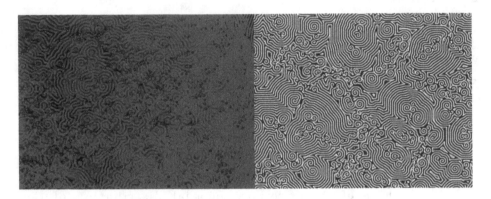

Fig. 1. Comparison of the Belousov-Zhabotinsky experiment record in the blue camera channel (left) and its NetLogo simulation (right).

The calculation was performed on a canvas of 1000×1000 cells. With a finer grid, the non-idealities of the periodic boundary did not suppress the system's behavior. Namely, the calculation with only a few points of ignition did not lead to the formation of regular structures.

In this paper, results achieved at $maxState = 200$ and $g = 28$ are discussed. Their role in the model is described in the next article in this volume.

For comparison to the proposed simulation, the B-Z reaction was performed as published previously [12].

3   Results

3.1   Traveling Through the Basin of Attraction

In the B-Z reaction, instead of observing the formation of a stable Turing pattern [13], we saw a regular stepwise interchange of self-similar structures which are in fact cellular automata, i.e., discrete dynamic networks [9]. In these systems, Gardens of Eden – configurations which the system may not reach from any previous, intermediate configurations – are sought. The next step in the trajectory consists of configurations that have a parental configuration. Many different Gardens of Eden often lead to one common structure, and there may be a few such "common" configurations on the trajectory towards the limit set. It is the state that a dynamical system reaches after an infinite amount of time. The limit set is where the system exhibits ergodic behavior. In discrete dynamic networks, it is either a fixed point or a periodically interchanging set of configurations.

In a continuous system in the two-dimensional space (e.g., plane), the properties of the limit set are governed by the Poincare-Bendixson Theorem [14]. It states that if a differentiable real dynamical system is defined on an open subset of the (two-dimensional) plane, then every non-empty compact ω-limit set of an orbit is either a fixed point, a periodic orbit, or a set composed of a finite number of fixed points connected by homoclinic and heteroclinic orbits. In higher dimensions, there may be other types of behavior, such as chaotic behavior or strange attractor. In discrete dynamic systems, chaotic behavior can arise in two or even one-dimensional systems.

In the final state of both the hodgepodge machine simulation and the B-Z experiment (Fig. 1), we observed a state which can be characterized as complex, i.e., having medium entropy and high variance. It is unclear how we can apply this terminology to the realm of multilevel cellular automata with more complicated rules, because the field is much less developed. However, we can ensure that it is not any type of periodic behavior predicted by the Poincare-Bendixson Theorem, because discrete systems typically display spirals [15].

3.2   Influence of the k_1 and k_2 Variation

In this section, results achieved at $maxState = 200$ and $g = 28$ are discussed. The role of $maxState$ and g ratio is described later in the text.

Figures 2, 3, 4, 5 and 6 show two possible trajectories of the hodgepodge machine to the basin of attraction. When $k_1 < 3$ or $k_2 < 3$, the initial concentric waves are circular. It is enabled by the production of waves from a single point, because $round(\frac{1}{2} + \frac{0}{2}) > 0$, $round(\frac{1}{2} + \frac{1}{2}) > 0$, and $round(\frac{0}{2} + \frac{1}{2}) > 0$. In contrast, if $k_1 \leq 3$ or $k_2 \leq 3$, we need at least one non-zero cell in the neighborhood of the zero cell to achieve the same effect, e.g., $round(\frac{1}{3} + \frac{1}{3}) > 0$. This subtlety has a very strong influence on the system's behavior.

In the early phase of the simulation (Fig. 2), for $k_1 = 2$ and $k_2 = 2$, we observed evolution of squares whose center appears to be circular. In contrast, the case $k_1 = 3$ and $k_2 = 3$ produced octagons with four long and four short sides.

Both cases of step 200 (Fig. 3) created a system of evolved waves. We observed a second type of the structure in the centre of the travelling wave system. For $k_1 = 3$ and $k_2 = 3$, we already observed a spiral doublet.

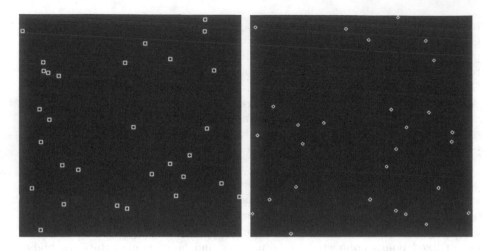

Fig. 2. The modified hodgepodge simulation in step 15 for $k_1 = 2$ and $k_2 = 2$ (left) and $k_1 = 3$ and $k_2 = 3$ (right).

In step 700 (Fig. 4), different k values continued to exhibit different behaviors. Spirals with a new type of wave emanating from the centre were "more stable" than the filamentous structures that appeared in both cases in Fig. 3).

In step 1200 (Fig. 5), the early wave structures vanished. For $k_1 = 2$ and $k_2 = 2$, the image was full of thin filamentous structures, whereas for $k_1 = 3$ and $k_2 = 3$, the image was almost identical to the final structure of the limit set.

Finally, in step 2000 (Fig. 6), for $k_1 = 3$ and $k_2 = 3$, the simulated structure fully matched the experiment structure. The ratio of spirals, wave

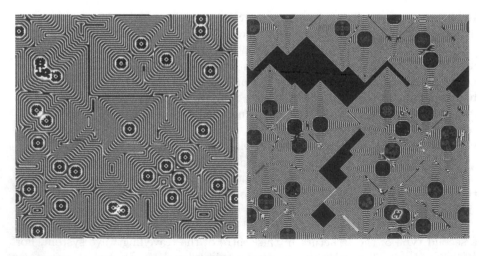

Fig. 3. The modified hodgepodge simulation in step 200 for $k_1 = 2$ and $k_2 = 2$ (left) and $k_1 = 3$ and $k_2 = 3$ (right).

structure fragments, and intermediate objects was stable within a certain limit. The case $k_1 = 2$ and $k_2 = 2$ was full of interchanging filamentous structures, which would, in further simulation, undergo a slow broadening into homogeneity (never observed in simulations). It seems that the limit set in the latter case is a certain type of coexistence of broad filaments. Since we do not have any experimental parallel to the filamentous type of structures, we continue with the analysis of the spiral structure.

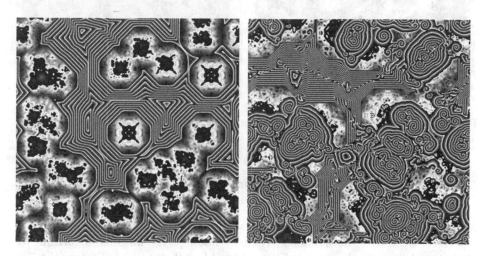

Fig. 4. The modified hodgepodge simulation in step 700 for $k_1 = 2$ and $k_2 = 2$ (left) and $k_1 = 3$ and $k_2 = 3$ (right).

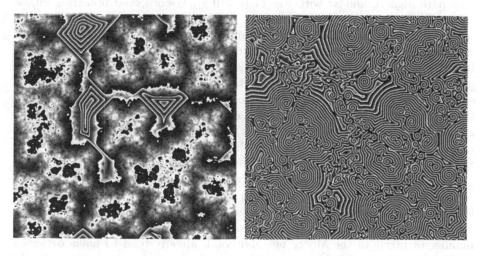

Fig. 5. The modified hodgepodge simulation in step 1200 for $k_1 = 2$ and $k_2 = 2$ (left) and $k_1 = 3$ and $k_2 = 3$ (right).

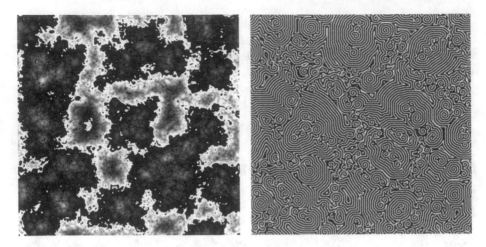

Fig. 6. The modified hodgepodge simulation in step 2000 for $k_1 = 2$ and $k_2 = 2$ (left) and $k_1 = 3$ and $k_2 = 3$ (right).

3.3 Influence of the Maximal Number of Cell States *maxState* on the Limit Sets

Multilevel cellular automata are often analyzed for at most 8 cell states, e.g. [9]. Here, we tested the modified Wilensky hodgepodge machine algorithm by changing the maximal number of the achievable cell states *maxStates* (Fig. 7).

The limit set – a mixture of waves and spirals – highly similar to those observed in the B-Z experiment was achieved at the $g/maxState$ ratio of 28:200 (compare Figs. 7c and 8c with Fig. 11 in 210 s). We observed branched spirals there (Fig. 7c). At the same ratio and higher *maxState* value (Fig. 7d), the spirals were even more smooth. To obtain these results, we kept the ratio constant and simplified it to 7:50. The simulation at the $g/maxState$ ratio of 3:20 brought up the limit set, which was also similar to the chemical experiment. The $g/maxState$ ratio of 1:7 (Fig. 7a) led to the formation of square spirals reported earlier [9,15]. The next ratio 28:200 (close to 1:7) gave results most similar to the experiment. The $g/maxState$ ratio of 1:8 (Fig. 7b) already showed spirals and waves like the B-Z experiment. For all comonnly studied "small-number" ratios, the simulation and reaction in the limit sets differ. At $maxState < 20$ the spiral evolutions were already qualitatively similar.

Among the low-number *maxState* levels, the simulation at the 1:7 ratio was extraordinary due to slow evolution to the limit set. Reaching the limit set took much higher number of simulation steps (37,000) than in other cases (e.g., 2,500 steps at $g/maxState = 1 : 8$). Indeed, 8 levels, which correspond to the number of pixels in the Moore neighborhood, already tend to make octagons (read below). At one level below (at 7 levels), the system's behavior was already changed. However, further research needs to be done to understand this phenomenon.

Fig. 7. The influence of the number of cell states *maxState* on the limit sets at similar *g/maxState* ratios of 1:7 (**a**), 3:20 (**b**), 28:200 (**c**), and 280:2000 (**d**). Total overviews of the limit sets' structures.

Fig. 8. The influence of the number of cell states *maxState* on limit sets at similar *g/maxState* ratios of 1:6 (**a**), 1:7 (**b**), 28:200 (**c**), and 280:2000 (**d**). Detailed views of the limit sets' structures.

3.4   Temporarily Organized Structures in the Initial Phases and the Limit Sets at Different $g/maxState$ Ratios

The temporarily organized structures on the state-space trajectory of the simulation are those on one of the paths of the discrete dynamic network [9]. In Figs. 9 and 10 we show the shape dependence of the temporarily organized structures in the initial phases and limit sets on the g value at the constant $maxState$ value (i.e., on the decreasing $g/maxState$ ratio).

Various ignition points, which are assumed to be Gardens of Eden, gave octagonal structures (Fig. 9), whose interiors were typical for the given Gardens of Eden. At the high $g/maxState$ ratio, after the passage of early waves when the

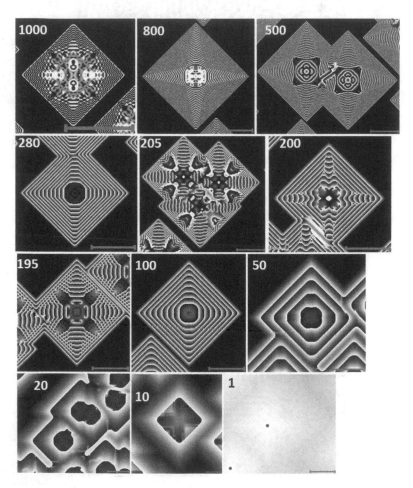

Fig. 9. The initial states for selected values of g at the constant $maxState = 2000$. Each of the scale bars corresponds to 100 px. As in the previous figures, the black color corresponds to the state level 0, white color to the maximal state value (set by the constant $maxState$).

interior octagon became regular, a centrally symmetrical structures appeared in the centre. The shapes of such initial central structures are dependent on the value of the $g/maxState$ ratio. We do not describe here the mechanism how these structures arise.

The early square waves gradually broadened with decreasing $g/maxState$ ratio (Fig. 9). The lower the $g/maxState$, the more diffuse the structures appeared. At $g/maxState \leq 50 : 2000$ we obtained a diffusive central object surrounded by the spreading wavefronts. With $g/maxState$ of 10:2000, the fuzzy diffusive wavefronts were becoming more sparse. At $g/maxState = 1 : 2000$ the second wave stopped evolving and the circular dark central structure appeared inside the dense diffusion.

Thus, the $g/maxState$ ratio can explain the thickness of the early waves and determine the number of states achieved in one time interval. This suggests that the wave thickness and density can, to some extent, explain other phenomena such as the built-in local asymmetry of the space, which ignites all additional phenomena and determines the state space trajectory of the process to its limit sets.

Fig. 10. The limit sets for selected g at $maxState = 2000$. From the upper left to lower right image, the values of g are 1000, 800, 500, 280, 150, 100, 20, 10, and 1, respectively.

The limit sets (Fig. 10) first emerge from a fully spiral form at $g/maxState = 1000 : 2000$. As $g/maxState$ decrease down to 280:2000, the spirals more and more resemble those found in a natural chemical reaction. The waves were broadening and by $g/maxState = 150 : 2000$ mature spirals could no longer be observed. Later, the organized structures were broken into a diffusive mixture of darker and lighter stains. At the limit $g/maxState \to 0$, we are likely to obtained a uniform intensity. It shows the unsuitability of the reaction-diffusion model for the description of the B-Z reaction [6].

3.5   Evolution of the State Space Trajectory

The main goal of the hodgepodge machine simulation is to test its consistency with the course of the B-Z reaction (an experimental trajectory). The segments of the reaction are depicted in Fig. 11.

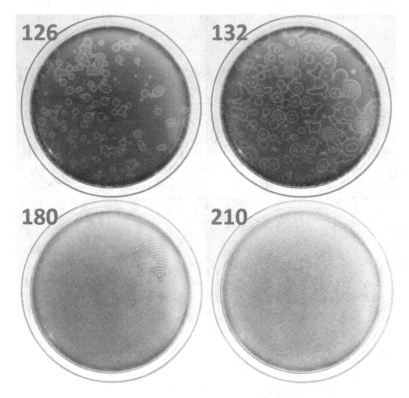

Fig. 11. Segments of the experimentally determined trajectory of the Belousov-Zhabotinsky reaction. Numbers correspond to the time interval (in seconds) from the initial mixing of the reactants.

Despite the similarity of the limit set between the simulation at $g/maxState$ ratio = 280:2000 (Figs. 7c and 8c) and the chemical experiment, the trajectories towards these limit sets were a bit different. In the case of the simulation (Fig. 12), after the ignition, the structures evolved in a sequence of square waves with round centers. After that, the first spiral doublet arose in the vicinity of two central objects such that the waves collided in one point. The doublet soon became surrounded by an elliptical wave thicker than the early square wave. This elliptical wave system had an expanding "diffusive" character until it was compressed by another system of elliptical waves which slowly evolved at other points.

In case of the model, early circular waves analogous to those in the experiment, were achieved only at very low $g/maxState$ ratios, when the waves were broad and the spirals were not formed (e.g., Fig. 9 for $g = 20$). This suggests that in the early phase of the experimental trajectory, there is a second process which brings the evolution of a different trajectory. Later, when this process is exhausted, the evolution of the system follows the path described by the hodge-podge machine model.

4   Discussion

The formation of spirals is somewhat mysterious. There currently does not exist a mathematical equation which describes the "robustness" or "stability" of the spiral. We also do not currently have a mathematical equation parallel to the Poincare-Bendixon Theorem for differentiable systems to describe our discrete system. Although, if it is a rule for discrete systems, why does it appear in the chemical systems in which we assume differentiability?

From the chemical point of view, the fundamental question in the interpretation of the B-Z reaction is whether the observed pattern is a result of a continuous behavior or whether the space is somehow discretised. Stable systems with two liquid phases – emulsions – are observed in many natural examples.

We propose that the similarity between the final phase of the experiment and the limit set of the simulation (Fig. 1) is not a coincidence. This suggests that the experimental problem may be discussed in the same way as the simulation. For the continuous case in the plane, where the Poincare-Bendixson Theorem holds, we should observe a fixed point, a limit cycle or a set of homoclinic and heteroclinic orbitals. However, this is not what we observed. Instead, we saw a mixture of wave fragments and spirals, where the spirals seem to be confined to discrete spatial arrangement. We conclude that, beyond the reasonable doubts, the structures observed in the B-Z reaction are the consequences of a certain kind of the space discretisation.

The fact that spirals are observed only in the case where two non-zero pixels exist in the neighborhood indicates that there must be a certain kind of local spatial asymmetry needed for spiral formation. Detailed discussion of this phenomenon is rather extensive and will be addressed later.

The time-spatial discretization is the idealized process, where the time element determines (a) a set of events achieved at the time within a spatial element (a cell) – the g constant – and (b) the rest of events on the spatial element's

184   D. Štys et al.

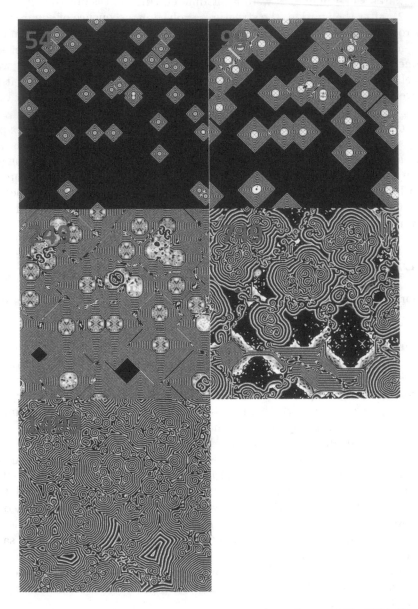

Fig. 12. The phases of the state-space trajectory at $g/maxState = 280 : 2000$. **Step 54** – The octagonal waves in the centre collapses and is replaced by the fractal structure. **Step 99** – In the places of the collision of two central structures, early spiral structures arise. **Step 249** – Spiral structures are being surrounded by the elliptical wavefront. **Step 749** – The simulation canvas is being filled by new spiral waves. **Step 1499** – Structures become almost regular by "compression" of waves of various origins.

(cell's) borders – the $maxState$ constant. We may consider analogies for such behaviour in known energy transfer processes, i.e. the difference between resonance energy transfer and energy transfer in systems of overlapping orbitals.

The analysis of these constants showed that the $g/maxState$ ratio determined the thickness of waves and the course of the state space trajectory irrespective to the total $maxState$ value. The decreasing $g/maxState$ ratio led to the loss of the wave structure, which eventually resulted in a fully diffusive picture. There also exists a lower limit of $maxState$ for the appearance successful course of the simulation. The higher the $maxState$ value, the smoother the spirals were.

Acknowledgments. This work was financially supported by CENAKVA (No. CZ.1.05/2.1.00/01.0024), CENAKVA II (No. LO1205 under the NPU I program) and The CENAKVA Centre Development (No. CZ.1.05/2.1.00/19.0380). Authors thank to Petr Jizba and Jaroslav Hlinka for important discussions and to Kaijia Tian for edits.

References

1. Belousov, B.P.: Collection of Short Papers on Radiation Medicine, p. 145. Medical Publications, Moscow (1959)
2. Zhabotinsky, A.M.: Periodical process of oxidation of malonic acid solution (a study of the Belousov reaction kinetics). Biofizika **9**, 306–311 (1964)
3. Zhabotinsky, A.M.: Periodic liquid phase reactions. Proc. Ac. Sci. USSR **157**, 392–395 (1964)
4. Rovinsky, A.B., Zhabotinsky, A.M.: Mechanism and mathematical model of the oscillating bromate-ferroin-bromomalonic acid reaction. J. Phys. Chem. **88**, 6081–6084 (1984)
5. Hagelstein, M., Liu, T., Mangold, S., Bauer, M.: Time-resolved combined XAS and UV-Vis applied to the study of the cerium catalysed BZ reaction. J. Phys. Conf. Ser. **430**, 012123 (2013)
6. Vanag, V.K., Epstein, I.R.: Cross-diffusion and pattern formation in reaction-diffusion systems. Phys. Chem. Chem. Phys. **11**, 897–912 (2009)
7. Gerhardt, M., Schuster, H.: A cellular automaton describing the formation of spatially ordered structures in chemical systems. Phys. D **36**, 209–222 (1989)
8. Dewdney, A.K.: The hodgepodge machine makes waves. Sci. Am. **225**, 104 (1988)
9. Wuensche, A.: Exploring Discrete Dynamics. Luniver Press, Frome (2011)
10. Wilensky, U.: NetLogo B-Z Reaction Model. Center for Connected Learning and Computer-Based Modeling, Northwestern Institute on Complex Systems, Northwestern University, Evanston, IL (2003). http://ccl.northwestern.edu/netlogo/models/B-ZReaction
11. http://www.biowes.org
12. Zhyrova, A., Stys, D.: Construction of the phenomenological model of Belousov-Zhabotinsky reaction state trajectory. Int. J. Comput. Math. **91**(1), 4–13 (2014)
13. Turing, A.M.: The chemical basis of morphogenesis. Philos. Trans. Roy. Soc. London **237**, 37–72 (1952)
14. Bendixson, I.: Sur les courbes définies par des équations différentielles. Acta Math. **24**, 1–88 (1901)
15. Fisch, R., Gravner, J., Griffeath, D.: Threshold-range scaling of excitable cellular automata. Stat. Comput. **1**, 23–39 (1991)

Parameter Identification Problem Based on FRAP Images: From Data Processing to Optimal Design of Photobleaching Experiments

Ctirad Matonoha[1]($\boxtimes$) and Štěpán Papáček[2]

[1] Institute of Computer Science, Academy of Sciences of the Czech Republic,
Pod Vodárenskou věží 2, 182 07 Prague 8, Czech Republic
matonoha@cs.cas.cz
[2] Faculty of Fisheries and Protection of Waters,
South Bohemian Research Center of Aquaculture and Biodiversity of Hydrocenoses,
Institute of Complex Systems, University of South Bohemia in České Budějovice,
Zámek 136, 373 33 Nové Hrady, Czech Republic
spapacek@frov.jcu.cz

Abstract. The aim of this study is to make a step towards optimal design of photobleaching experiments. The photobleaching techniques, mainly FRAP (Fluorescence Recovery After Photobleaching), are widely used since 1970's to determine the mobility of fluorescent molecules within the living cells. While many rather empirical recommendations for the experimental setup have been made in past decades, no rigorous mathematical study concerning optimal design of FRAP experiments exists. In this paper, we formulate and solve the inverse problem of data processing of FRAP images leading to the underlaying model parameter identification. The key concept relies on the analysis of sensitivity of the measured outputs on the model parameters. It permits to represent the resulting parameter estimate as random variable, i.e., we can provide both the mean value and standard error or corresponding confidence interval. Based on the same sensitivity-based approach we further optimize experimental design factors, e.g., the radius of bleach spot. The reliability of our new approach is shown on a numerical example.

Keywords: FRAP · Optimal experimental design · Sensitivity analysis · Parameter identification

1 Introduction

The image processing is certainly one of the fastest growing areas in informatics and applied mathematics. Many new applications, e.g., in biology and medicine, rise up every year. However, there is a gap between the level of sophistication of equipment for the data acquisition and the quality of further data processing. Particularly, discussion about the data noise propagation (from data to

© Springer International Publishing Switzerland 2016
T. Kozubek et al. (Eds.): HPCSE 2015, LNCS 9611, pp. 186–195, 2016.
DOI: 10.1007/978-3-319-40361-8_14

the resulting parameter estimates), i.e., the error or uncertainty analysis corresponding to respective methods, is rare and whole concept of parameter definition as random variable is often misunderstood by the biological community, cf. [9,10,15].

While in our previous papers we sought to elaborate reliable methods for the processing of spatio-temporal images acquired by the so-called FRAP (Fluorescence Recovery After Photobleaching) method [6,7,11–13], the aim of the present study is to make a step from the data processing to optimal design of photobleaching experiments. Further we show how to find a specific "optimal" experimental conditions maximizing a measure of sensitivity defined as the sum of squares of partial derivatives of the measured output on the estimated parameter, cf. (11).

Both FRAP & FLIP (Fluorescence Loss In Photobleaching) are based on the measuring the change in fluorescence intensity in a region of interest, e.g., in a finite 2D domain representing the part of a membrane, in response to an external stimulus (bleaching). A high-intensity laser pulse provided by the confocal laser scanning microscopy (CLSM) causes a presumably irreversible loss in fluorescence in the bleached area and the subsequent recovery in fluorescence reflects the mobility (related to diffusion) of fluorescent compounds from the area outside the bleach spot. CLSM allows to obtain high-resolution optical images with deep selectivity, however, the small energy level emitted by the fluorophore and the amplification performed by the photon detector introduces a measurement noise making the subsequent parameter identification problem highly unstable due to the ill-posedness in Hadamard's sense [3,4].

The rest of this paper is organized as follows: In Sect. 2 we describe the FRAP & FLIP data acquisition and structure. In Sect. 3 we formulate the inverse problem of parameter identification and introduce the sensitivity analysis. Then, in Sect. 4, we develop a new theoretical approach allowing the optimization of FRAP experimental factors and provide one numerical example. The novelty of our approach and outlooks for further research are discussed in the final Sect. 5.

2 Data Acquisition and Data Structure

The spatio-temporal FRAP data are graphically depicted in Fig. 1. Usually, the images are made with certain time period (in our case every 8s) before and after the application of high-intensity laser pulse (so-called bleach). The pre-bleach image (see the top left image in Fig. 1) shows a typical distribution of phycobilisome fluorescence in a single cell. Application of high laser intensity across the vertical axis (red rectangle in the second image in top row in Fig. 1) reduced phycobilisome fluorescence to about 40 % of the initial value due to the destruction of a portion of the phycobilin pigments. The observed fluorescence recovery in the bleached zone is attributed to phycobilisome mobility in this red alga, see e.g., [5] and references therein.

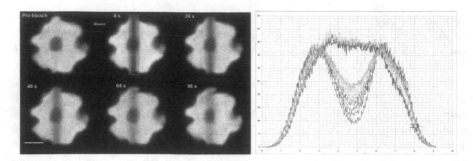

Fig. 1. Left: representative FRAP image sequence for a single cell of red algae *Porphyridium cruentum* for phycobilisome fluorescence. First, a fluorescence image before bleaching was detected (pre-bleach), then the phycobilisome fluorescence was bleached out across the middle of the cell in the vertical direction (red dashed rectangle). The sequence of five post-bleach images is shown. The length of the scale bar is 3 μm. Right: experimental (noisy) data in form of one-dimensional bleach profiles for different time instants after the bleach. The abscissa represents the position along the axis perpendicular to the bleach stripe. In the ordinate there is the corresponding average fluorescence (in arbitrary units) along the axis parallel to bleach. In the central region we see the step-wise recovery of the signal: from the lowest value (first post-bleach) to the highest pre-bleach (steady-state) values on the top (Color figure online).

A FRAP data structure usually consists of a time sequence of rectangular matrices, where each entry quantifies the fluorescence intensity u at a particular spatial point in a finite 2D domain (e.g., by a number between 0 and 255):

$$u(x_{kl}, t_j)_{j=0}^{N_t}, \quad k = 1 \dots N_x, \ l = 1 \dots N_y,$$

where k, l are the spatial indexes uniquely identifying the pixel position where the signal u is measured, and j is the time index (the initial condition corresponds to $j = 0$), cf. [6,7,11]. Usually, the data points are uniformly distributed both in time (the time interval Δt between two consecutive measurements is constant) and space, i.e., on an equidistant 1D or 2D mesh. Let see the right part of Fig. 1, where we observe an example of 1D fluorescence intensity profiles (in arbitrary units) for different time instants $t_0 \dots t_{N_t}$.

Further, in sake of simplicity, we shall infer about the parameter D by using direct measurements of discrete data in a space-time domain when only one index is employed, i.e., we use the following form of data

$$u(x_i, t_i)_{i=1}^{N_{\text{data}}} \in \mathbb{R}^{N_{\text{data}}}.$$

3   Problem Formulation

Let us consider the isotropic diffusion process characterized by one single scalar parameter: a diffusion coefficient D (constant in space). Right now we assume that D is time-dependent, i.e., an anomalous diffusion is allowed. The governing equation for the spatio-temporal fluorescence signal $u(x,t)$, proportional to the fluorescent particles concentration, is Fick's diffusion equation as follows

$$\frac{\partial}{\partial t}u(x,t) = D\Delta u(x,t) \qquad\qquad x \in \Omega, \ t \in [0,T] \qquad\qquad (1)$$

$$u(x,0) = u_0(x) \qquad\qquad x \in \Omega \qquad\qquad (2)$$

$$\text{boundary conditions} \qquad\qquad \partial\Omega \times [0,T]. \qquad\qquad (3)$$

Boundary conditions could be, e.g.,

$$u(x,t) = 0 \quad \text{or} \quad \frac{\partial}{\partial n}u(x,t) = 0 \quad \text{on} \quad \partial\Omega \times [0,T].$$

We also consider the simplest case of unbounded domains $\Omega = \mathbb{R}^n$, in which case we set appropriate decay conditions at $|x| \to \infty$, $t \in [0,T]$. The above formulation (1)–(3) (and variants) is the basis for all the further analysis.

In the case of constant coefficient D, the solution to this problem can be expressed by means of the Green function $G(x,t;y)$ for the heat equation

$$\frac{\partial}{\partial t}G(x,t;y) = \Delta G(x,t;y) \qquad\qquad x \in \Omega, \ t \in [0,T]$$

$$G(x,0;y) = \delta(x-y) \qquad\qquad x \in \Omega$$

$$\text{boundary conditions for } G(x,t;y) \qquad\qquad \partial\Omega \times [0,T].$$

Some frequently used cases are that of a diffusion in free space, e.g., in the one-dimensional domain $\mathbb{R}$ without boundary conditions, the Green function is the heat kernel

$$G(x,t;y) = \frac{1}{\sqrt{4\pi t}} \exp\left[-\frac{(x-y)^2}{4t}\right] \qquad\qquad x,y \in \mathbb{R}.$$

In FRAP experiments, the initial condition, i.e., the first post-bleach profile (with the background or pre-bleach signal subtracted) is often modeled as a Gaussian, cf. Fig. 1, which leads in the one-dimensional case to initial condition of the form

$$u_0(x) = u_{0,0} \exp\left(-\frac{2x^2}{r_0^2}\right), \qquad\qquad (4)$$

where $u_{0,0} \geq 0$ is the maximum depth at time t_0 for $x = 0$, $r_0 > 0$ is the half-width of the bleach at normalized height (depth) $\exp(-2)$, i.e., $\frac{u_0(r_0)}{u_{0,0}} = \exp(-2)$, cf. [11]. An explicit solution for u in the one-dimensional free space case is then given by

$$u(x,t) = u_{0,0}\frac{r_0}{\sqrt{r_0^2 + 8Dt}} \exp\left(-\frac{2x^2}{r_0^2 + 8Dt}\right). \qquad\qquad (5)$$

Parameter Identification Problem Based on FRAP data

Define a forward map (also called a parameter-to-data map)

$$F : \mathbb{R} \to \mathbb{R}^{N_{\mathrm{data}}} \tag{6}$$

$$(D) \to u(x_i, t_i)_{i=1}^{N_{\mathrm{data}}}. \tag{7}$$

Our regression model is now

$$F(D) = \mathrm{data}, \tag{8}$$

where the data are modeled as contaminated with additive white noise

$$\mathrm{data} = F(D_T) + e = u(x_i, t_i)_{i=1}^{N_{\mathrm{data}}} + (e_i)_{i=1}^{N_{\mathrm{data}}}. \tag{9}$$

Here D_T denotes the true coefficient and e is a data error vector which we assume to be normally distributed with variance σ^2

$$(e_i)_{i=1}^{N_{\mathrm{data}}} \in \mathbb{R}^{N_{\mathrm{data}}}, \quad e_j = \mathcal{N}(0, \sigma^2), \quad j = 0, \dots, N_t.$$

Given some data, the aim of the parameter identification problem is to find D such that (8) is satisfied in some appropriate sense. Since (8) usually consists of an overdetermined system (there are more data points than unknowns), it cannot be expected that (8) holds with equality, but instead an appropriate notion of solution (which we adopt for the rest of the paper) is that of a least-squares solution D_c (with $\|.\|$ denoting the Euclidean norm on $\mathbb{R}^{N_{\mathrm{data}}}$):

$$\|F(D_c) - \mathrm{data}\|^2 = \min_{D>0} \|F(D) - \mathrm{data}\|^2. \tag{10}$$

The above defined parameter identification problem is usually ill-posed for nonconstant coefficients, so that regularization has to be employed; see, e.g., [4]. A solution of practical example based on FRAP data was presented in [11].

Sensitivity Analysis and Confidence Intervals

For the sensitivity analysis, cf. [2, 7], we require the Fréchet-derivative $F'(D) \in \mathbb{R}^{N_{\mathrm{data}} \times 1}$ of the forward map F, that is

$$F'(D) = \frac{\partial}{\partial D} F(D) = \begin{pmatrix} \frac{\partial}{\partial D} u(x_1, t_1) \\ \cdots \\ \cdots \\ \frac{\partial}{\partial D} u(x_{N_{\mathrm{data}}}, t_{N_{\mathrm{data}}}) \end{pmatrix}.$$

A corresponding quantity used further as our key sensitivity measure is a number

$$M(D) = F'(D)^T F'(D) \quad \in \mathbb{R}. \tag{11}$$

Based on the book of Bates and Watts [1], we can estimate confidence intervals. Suppose we have computed D_c as least-squares solutions in the sense of (10). Let us define the residual as

$$res^2(D_c) = \|F(D_c) - \text{data}\|^2 = \sum_{i=1}^{N_{\text{data}}} [\text{data}_i - u_{D_c}(x_i, t_i)]^2, \qquad (12)$$

where u_{D_c} is computed from (1)–(3) for the parameter value D_c. Then according to [1], it is possible to quantify the error between computed parameter D_c and true parameter D_T. In fact, we have an approximate $1 - \alpha$ confidence interval

$$(D_c - D_T)^2 \sum_{i=1}^{N_{\text{data}}} \left[\frac{\partial}{\partial D} u(x_i, t_i)\right]^2 \leq \frac{res^2(D_c)}{N_{\text{data}} - 1} f_{1, N_{\text{data}} - 1}(\alpha). \qquad (13)$$

In equation (13), several simplifications are possible. Note that according to our noise model, the residual term $\frac{res^2(D_c)}{N_{\text{data}} - 1}$ is an estimator of error variance such that an approximation

$$\frac{res^2(D_c)}{N_{\text{data}} - 1} \sim \sigma^2 \qquad (14)$$

holds for N_{data} being large [1]. The term $\frac{res^2(D_c)}{N_{\text{data}} - 1}$ in (13) can be viewed as rather independent of D_c or N_{data}. Moreover, we remember the reader that the Fisher distribution with 1 and $N_{\text{data}} - 1$ degrees of freedom converges to the χ^2-distribution as $N_{\text{data}} \to \infty$. Hence, the term $f_{1, N_{\text{data}} - 1}(\alpha)$ can approximately be viewed as independent of N_{data} as well and of moderate size.

4   Optimizing Experimental Design Variables

There are many rather *empirical recommendations* related to the design of a photobleaching experiment, e.g., the bleach spot shape and size (design factor r_0), the region of interest location and size (design factor L), total time of measurement (T), see [7,15] and references therein. However, we should have a *more rigorous tool* for the choice of experimental design factors. Based on the process model (1)–(3) and just introduced sensitivity analysis, we can define an optimization problem residing in the maximization of the sensitivity measure (11).

The key parameter in FRAP measurements is the size (and shape) of bleach spot, e.g., the characteristic radius r_0 in case of a circular bleach. If the size of bleach spot can be varied (at the same time keeping the bleach depth $u_{0,0}$ fixed), we should ask the question if there is an optimal bleach size that can be used. Thus, we can try to look for such a bleach radius r_0 which leads to maximal sensitivity since this corresponds to minimal confidence intervals (for comparable experiments).

More precisely, in the one-dimensional case of the Fick diffusion on a line, having the set of observations on a space-time cylinder $Q = [-L, L] \times [0, T]$, we try to infer about the optimal bleach radius r_{opt} yielding maximal sensitivity.

We introduce a function

$$S(r_0) = \sum_{i=1}^{N_{\text{data}}} \left[\frac{\partial}{\partial D} u(x_i, t_i) \right]^2 = \sum_{k=1}^{N_x} \sum_{j=1}^{N_t} \left[\frac{\partial}{\partial D} u(x_k, t_j) \right]^2, \qquad (15)$$

where $N_x = \frac{2L}{\Delta x} + 1$ and $N_t = \frac{T}{\Delta t}$, and we try to find out a maximal value

$$S(r_{opt}) = \max_{r_0 > 0} S(r_0).$$

Note that $S(r_0)$ is equal to M from (11).

Numerical Example

In the following example we compute a least-squares estimate D_c and the sensitivity $S(r_0)$, cf. (15). We consider a rectangular spatio-temporal data grid with space interval $x_i \in [-6, 6]$, i.e., $L = 6$, and time interval $t_i \in [0, T]$ for various T. For our test purposes we used various grid sizes Δx and Δt and also various exact diffusion coefficient D_T. We simulated data by assuming D_T with different bleach radii r_0 and computed the data for the 1D case by (5). Based on these data we computed a least-squares estimate D_c of the diffusion coefficient using a procedure described in [11]. It is a one-dimensional minimization problem (10) for D. To obtain a solution, we used variable metric method implemented in our optimization system [8]. The values $M = S(r_0)$ were then computed numerically using central differences.

To see what may influence a value of optimal bleach radius r_{opt}, we considered different values of $D_T, T, \Delta x, \Delta t$ defined in Table 1.

Table 1. Input values for numerical experiments.

Data set	D_T	Δx	N_x	Δt	N_t	T
Data 1	1	0.1	121	0.1	40	4
Data 2	1	0.1	121	0.01	400	4
Data 3	1	0.01	1201	0.1	40	4
Data 4	2	0.1	121	0.01	200	2
Data 5	1	0.1	121	0.01	200	2
Data 6	2	0.1	121	0.01	100	1

Typical behaviors of dependence of $M = S(r_0)$ on r_0 and computed values D_c on r_0 are shown in Fig. 2. For this purpose we used the results for data set *Data 1*. One can see that there exists a unique maximum of function $S(r_0)$ which is marked with a black circle. There exists an optimal bleach radius r_{opt} leading to maximal sensitivity.

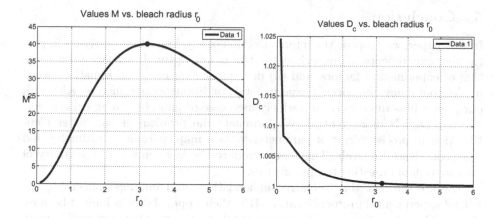

Fig. 2. Values $M = S(r_0)$ and D_c vs. bleach radius r_0 for data set *Data 1*.

Optimal bleach radii for all data sets together with computed values D_c are presented in Table 2. We found out that the optimal bleach radius is the same for data sets *Data 1 – Data 4* and for data sets *Data 5 – Data 6*. The value of r_{opt} is influenced by exact diffusion coefficient D_T and time interval of measurement T. The function value $S(r_{opt})$ depends on the number of spatio-temporal points. For example, this value is approximately 10 times larger for data sets *Data 2* and *Data 3* in comparison with data set *Data 1* because the number of points is 10 times larger (10 times larger number of $N_x N_t$), see the sums in (15).

Table 2. Results of numerical experiments.

Data set	r_{opt}	$S(r_{opt})$	D_c	$|D_c - D_T|$
Data 1	3.2	40.13	1.000528	5.28E–4
Data 2	3.2	395.36	1.000343	3.43E–4
Data 3	3.2	399.77	0.999999	1.00E–6
Data 4	3.2	198.08	2.001091	1.09E–3
Data 5	2.4	149.83	1.000810	8.10E–4
Data 6	2.4	75.21	2.001912	1.91E–3

The obtained results correspond quite well with our theoretical findings published in [14], where we argue that the value r_{opt} depends on the square root of the product of time interval of measurement T and exact diffusion coefficient D_T. Indeed, the optimal value r_{opt} is the same for the same product $T D_T$.

5 Conclusion

In this paper, we propose the interconnection of two important activities in performing experiments: (i) experimental design, i.e., optimal or near-optimal setting of experimental factors, and (ii) data processing based on a mathematical model containing the specific experimental conditions as parameters. Although our idea is illustrated only on a widely used case of photobleaching experiment, our approach is more general. We formulate the problem of parameter identification in precise terms of parameter-to-data map, parameter estimates and their confidence intervals. Then, we introduce the key concept of sensitivity of measured data on estimated parameters.

Despite the fact that some recommendations and findings concerning the FRAP experimental protocol exist, cf. [10], their applicability is limited because they are based on very specific experimental conditions. Our approach is more general and accurate (always when the process model is reliable).

In order to validate our idea of the model-based optimization of experimental conditions, we provide one numerical example. We prove that one of the most important experimental design factors in photobleaching experiments, the bleach size r_0, can be actually optimized, i.e., there exists a value r_{opt} for which the sensitivity measure $S(r_{opt})$ reaches the maximal value, hence assuring the shortest confidence interval, cf. (13).

Our findings are expected to be incorporated into a process of FRAP experimental protocol development – it is not computationally expensive and the enhancement of the parameter estimation process can be substantial, e.g., a four times higher $S(r)$ assures half upper bound for the standard error of the estimated parameter, cf. Fig. 2.

Certainly, the more realistic model formulation should be conceived in order to get reliable results, e.g., taking into account the anisotropic diffusion on finite 2-dimensional domain, binding reaction, bleaching during scanning, more general bleaching shapes and topologies. All these issues are only some extension of the presented study and do not question neither the governing Fick diffusion PDE nor the nature of the computation domain Ω (if it is a Euclidean domain or a fractal set modelling the molecular crowding). This is the subject of our ongoing research together with an ambitious goal consisting of the computationally effective *on-line* model-based sensitivity analysis. The appealing idea is to suggest the optimal values of experimental design variables *on-line*, i.e., to perform the experimental protocol modification (or tuning) during FRAP measurements. The main drawback of this very last idea is neither mathematical nor technical difficulty but the complicated communication between the members of mathematical and biological community.

Acknowledgement. This work was supported by the long-term strategic development financing of the Institute of Computer Science (RVO:67985807) and by the Ministry of Education, Youth and Sport of the Czech Republic? projects 'CENAKVA' (No. CZ.1.05/2.1.00/01.0024) and 'CENAKVA II' (No. LO1205 under the NPU I program). SP thanks to Stefan Kindermann (JKU Linz, Austria) for inspiring discussions.

References

1. Bates, D.M., Watts, D.G.: Nonlinear Regression Analysis: Its Applications. Wiley, New York (1988)
2. Cintrón-Arias, A., Banks, H.T., Capaldi, A., Lloyd, A.L.: A sensitivity matrix based methodology for inverse problem formulation. J. Inverse Ill-Posed Prob. **17**, 545–564 (2009)
3. Hadamard, J.: Lectures on the Cauchy Problem in Linear Partial Differential Equations. Yale University Press, New Haven (1923)
4. Engl, H., Hanke, M., Neubauer, A.: Regularization of Ill-Posed Problems. Kluwer, Dortrecht (1996)
5. Kaňa, R., Kotabová, E., Lukeš, M., Papáček, Š., Matonoha, C., Liu, L.N., Prášil, O., Mullineaux, C.W.: Phycobilisome mobility and its role in the regulation of light harvesting in red algae. Plant Physiol. **165**, 1618–1631 (2014)
6. Kaňa, R., Matonoha, C., Papáček, Š., Soukup, J.: On estimation of diffusion coefficient based on spatio-temporal FRAP images: an inverse ill-posed problem. In: Chleboun, J., Segeth, K., Šístek, J., Vejchodský, T. (eds.) Programs and Algorithms of Numerical Mathematics 16, pp. 100–111 (2013)
7. Kindermann, S., Papáček, Š.: On data space selection and data processing for parameteridentification in a reaction-diffusion model based on FRAP experiments. Abstr. Appl. Anal. **2015**, Article ID 859849 (2015)
8. Lukšan, L., Tůma, M., Matonoha, C., Vlček, J., Ramešová, N., Šiška, M., Hartman, J.: UFO 2014 - interactive system for universal functional optimization. Technical report V-1218, Institute of Computer Science, Academy of Sciences of the Czech Republic, Prague (2014). http://www.cs.cas.cz/luksan/ufo.html
9. Mai, J., Trump, S., Ali, R., Schiltz, R.L., Hager, G., Hanke, T., Lehmann, I., Attinger, S.: Are assumptions about the model type necessary in reaction-diffusion modeling? FRAP Appl. Biophys. J. **100**(5), 1178–1188 (2011)
10. Mueller, F., Mazza, D., Stasevich, T.J., McNally, J.G.: FRAP and kinetic modeling in the analysis of nuclear protein dynamics: what do we really know? Curr. Opin. Cell Biol. **22**, 1–9 (2010)
11. Papáček, Š., Kaňa, R., Matonoha, C.: Estimation of diffusivity of phycobilisomes on thylakoid membrane based on spatio-temporal FRAP images. Math. Comput. Model. **57**, 1907–1912 (2013)
12. Papáček, Š., Jablonský, J., Matonoha, C.: On two methods for the parameter estimation problem with spatio-temporal FRAP data. In: Chleboun, J., Segeth, K., Šístek, J., Vejchodský, T. (eds.) Programs and Algorithms of Numerical Mathematics 17, pp. 100–111 (2015)
13. Papáček, Š., Jablonský, J., Matonoha, C., Kaňa, R., Kindermann, S.: FRAP & FLIP: two sides of the same coin? In: Ortuño, F., Rojas, I. (eds.) IWBBIO 2015, Part II. LNCS, vol. 9044, pp. 444–455. Springer, Heidelberg (2015)
14. Papáček, Štepán, Kindermann, Stefan: On optimization of FRAP experiments: model-based sensitivity analysis approach. In: Ortuño, F., Rojas, I., et al. (eds.) IWBBIO 2016. LNCS (LNBI), vol. 9656, pp. 545–556. Springer, Heidelberg (2016). doi:10.1007/978-3-319-31744-1_49
15. Sbalzarini, I.F.: Analysis, Modeling and Simulation of Diffusion Processes in Cell Biology. VDM Verlag Dr. Muller, Saarbrücken (2009)

Author Index

Printed in the United States
by Bookmasters

Printed in the United States
By Bookmasters